AF411868

SAFE AGUA BOOK DESIGN

MANAGING EDITOR
Liliana Becerra

EDITORIAL TEAM
Mariana Amatullo
Dan Gottlieb
Penny Herscovitch
Elisa Ruffino

EDITOR
Alex Carswell

ART DIRECTION
Lisa Wagner

SENIOR DESIGNER
Giancarlo Llacar

PRINTED BY
Typecraft, Wood and Jones
Pasadena, CA

FACULTY
Lisa Wagner,
Graphic Design

TEACHING ASSISTANT
K.C. Cho,
Product Design

STUDENT CREATIVE TEAM
Pei-Jeane Chen,
Graphic Design

Evangeline Joo,
Illustration

Karen Ko,
Graphic Design

Giancarlo Llacar,
Graphic Design

Ping Zhu,
Illustration

Eric Mathias,
Graphic Design

This publication is made possible thanks in part to the generous support of the Chilean Ministry of
Foreign Affairs, which is endorsing the Safe Agua Project within the context of the Chile-California Plan.

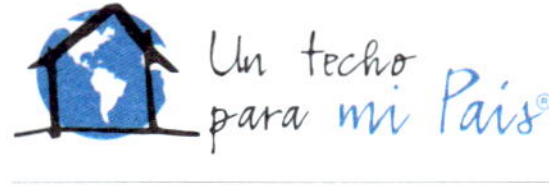

SAFE AGUA

WE DEDICATE THIS BOOK WITH GREAT APPRECIATION AND GRATITUDE TO ALL THE FAMILIES FROM CAMPAMENTO SAN JOSÉ, SANTIAGO, CHILE.

Designmatters I Art Center College of Design

THE BEGINNING

THE BIG PICTURE

OUR APPROACH

PROJECTS

EPILOGUE

Cactus Leaves / Once upon a time, Father Felipe Berríos was traveling in a remote region of Chile. He visited a village where the natives had hardly any clean water for their daily activities. He was standing by a nearby river, and noticed that the water was very dirty. As he watched, a native woman came down to the river to collect water in a bucket. When Father Felipe saw what she was doing, he warned her that the water was dirty and that she shouldn't use it for her daily chores. / Calmly, the woman sat down on the ground with the bucket full of dirty water that she had just collected. She took a leaf from a cactus that was growing nearby and cut it open sideways. She turned it inside out, so that she could put her hand inside, as if the leaf were a glove. / She then dipped her hand, with the cactus "glove" on, into the water and started to stir in a circular motion. Suddenly, thanks to the effect of the cactus leaf, all the dirt from the water started to come together, forming a small ball. Little by little, as the dirt ball grew larger, the water became completely transparent and clean. The dirt ball dropped to the bottom of the bucket and the woman reached down and took it out. She proudly showed the newly clear water to Father Felipe, and happily carried her full bucket back to the village.

By **Felipe Berríos**, Founder of *Un Techo para Chile*,
as told to **Liliana Becera**, Faculty, Product Design,
Art Center College of Design

A Life-Changing Experience

By Lorne M. Buchman

THE PROJECT SAFE AGUA addresses a basic necessity that is easily taken for granted by those of us who have it available at our fingertips with just the turn of a tap. In developing countries and impoverished communities around the world, obtaining clean water is a daily struggle, and every drop is understandably precious.

I can still remember a question that one of my teachers asked us in grade school: "How many new water molecules have been created on the earth during the last several million years?" The answer, of course, is none, and the fact that there is a finite, recycled supply of water in the world made a lasting impression on me. According to the US Geological Survey Web site, the fresh water needed for everyday use by humans comes primarily from rivers and lakes that contain only 0.0067% of the total amount of water on our planet. When you take all of this into consideration, it reinforces just how limited safe drinking water is as a resource. The problem of access to clean water is compounded further when you contemplate how unevenly it is distributed around the globe due to geography, climate, and socioeconomic factors.

Addressing the challenge of safe water access — in all its complexity — formed the basis of a brilliant educational studio project at Art Center College of Design in the summer of 2009. This publication is a tribute to, and a record of, its stunning achievement. Safe Agua took form as a collaboration between Art Center College of Design and *Un Techo para mi País* to address the everyday needs of the impoverished people in the Campamento San José, given their extremely limited access to potable water. The highly dedicated team of Art Center students and instructors were all deeply affected by their experience. The exploratory trip they took to Chile to meet individuals and families who are living under these difficult conditions, of which safe drinking water is only one of the many hardships faced, was life changing. It impressed upon them not only the tremendous challenge of this community's adverse circumstance, but also the fortitude, dignity, and blessed sense of humor with which the people there carried on living their lives. The trials of this community — how they use water, how they access water, how they might find solutions to real social, practical, and physical issues concerning water use — all became the focus of study for our students and faculty.

The Designmatters program at Art Center is significant in our curriculum and offers an exceptional opportunity for students to work on projects of humanitarian and social impact. It is projects like Safe Agua that clearly illustrate the influence design and design education can play in addressing significant real-world problems. There is also an invaluable lesson in the approach taken to the work of those involved; this was not a group thrusting ideas and solutions on a foreign culture. The conversation with community occurred first. The students lived there. They talked to people. They listened. The learning involved was important for them not only as design professionals but, equally important, as members of a global society who respect difference. Only with that sensibility would the solutions be real.

I want to celebrate the hard work and dedicated efforts of all the students and faculty, and the leadership of Designmatters, who have given so much of themselves to try and help those in need. I do not doubt that the experience they gained has made a lifelong impact on them, as well as on everyone they met in Campamento San José.

Dr. Lorne M. Buchman is President of Art Center College of Design.

A Curriculum for
Social Change

By Adlai Wertman

THERE IS CERTAINLY NO DEARTH of problems facing the world. Global poverty, environmental breakdowns, and health crises abound. The traditional models of international development have, in general, failed as they leave behind empty, crumbling schoolhouses and villages with infrastructure but no jobs. Traditional nonprofit models are also under enormous stress as they compete to raise money for ongoing social services. And while these NGOs are to be applauded for the amazing work they do, the results are often unsustainable, and risk creating long-term trends toward dependence.

At the same time, private or public for-profit companies are strapped by their need to produce next-quarter returns on investments. The profit motive forces a focus on markets where high margins are the rule and where the poor are typically written off. For most, this means ignoring the poorest people in the world — those we have come to describe as being at "the bottom of the pyramid." This group consists of 2.5 billion individuals living on under $3 per day. And although strategists believe that there must be a market there for value products, there are few examples of pure for-profit entities succeeding in raising the standard of living in poverty stricken countries.

These challenges have brought new disciplines to the sphere of attacking global problems. The medical and public health fields have long seen their ability to have deep impact in these regions. Engineers have also found many applications for their expertise. But two disciplines — design and business — had, until the relatively recent past, yet to find their roles.

For many, "design" conjures up images of people drawing pictures, designing products, and creating logos. However, this was never the true purpose and use of design. Designers look at how people live and create ways of making those lives better, easier, and more functional — as well as more appealing. When the term "design thinking" entered the boardrooms of the world, it instigated an approach that sought creative solutions in an empathetic and rational paradigm. As design thinking entered the business mainstream, it led to new products that made a great deal of money for companies

Only recently, though, have design and design thinking been brought to bear on solving the world's biggest problems. The questions Mariana Amatullo and her colleagues at Art Center asked were simple enough: Could design students use their training and creativity to help the poorest people on the earth? Could designers take on environmental challenges in Guatemala? Could they impact critical health issues from Africa to East Los Angeles? And, in the case at hand here, could they take on water challenges in Chile? In seeking answers to those and similar questions, Designmatters at Art Center was born. And the results have been nothing short of amazing.

The project chronicled in this book challenged students to use their design skills to address how people access, use, and conserve water in new developments in the poorest parts of Chile. As you will see, the products and processes they came up with were astounding and impactful.

When I had the privilege of meeting Mariana, I asked some important questions. Would this all be economically sustainable, or would we be repeating the same international development failures — dropping these new products and processes out of a helicopter, so to speak, and going back home? Or, alternatively, could we not only make these designs beneficial but also build a local economy around them?

This is where business comes into the picture. Over the past few decades a new breed of people came to the social sector. They were trained in business, but had a desire to use their education and experience to face off against the challenges of poverty, health, and the environment. These folks, dubbed "social entrepreneurs," were building business models to create sustainable change in the lives of the poorest people in the world.

One notable example of their success is micro-finance. Born in Southeast Asia, the model quickly swept across the world. Micro-finance banking institutions now offer micro-loans, micro-savings, and micro-insurance in a market now estimated to be in the hundreds of billions of dollars. And while most of these institutions may always need some form of philanthropic support — even if only for start-up — they are all creating self-sufficiency on the part of their clients.

So then, could business models be used to take the ideas and products coming out of Designmatters projects such as Safe Agua to create sustainable change? Could we build an economy around them? Could we build the brilliant shower device in this book using local materials, have them assembled locally and distributed throughout the region? The results would not only be the ability to shower with dignity, but also the creation of jobs in manufacturing, sales, distribution, and maintenance. The showers could be built in multiple towns and sold at value-driven prices beyond those immediate markets. Those towns would bring in much needed currency and build a market economy where there may only have been subsistence farming before.

The *Mila* community laundry facilities could become micro-franchises. Local groups of women could take micro-loans, buy the machines and take in laundry. Again, this would create jobs and income while freeing mothers to educate their children and find new ways to start their own businesses — perhaps making the new dishwashing product also designed by the students.

But creating these micro-economies requires business skills. The locals must learn how to price their goods and manage cash flow. Selling the products requires marketing plans and sales skills. Co-ops must be created and organized. Loans and insurance must flow. Creating economic sustainability therefore requires marrying designers with people trained in business. The partnership is a natural one, in which business people push designers to do more analytical thinking and designers teach business people the basics of design thinking.

The program I began at the Marshall School of Business at the University of Southern California — the Society and Business Lab — teaches undergraduate and graduate business students how to take part in the growing field of social entrepreneurship. When they saw what the Designmatters portfolio was accomplishing, their eyes lit up. It was like giving teachers the gift of a great curriculum; we were able to hand our business students products, services, and processes to work with.

I have great hopes for the future of our two programs. But moreover, I have great confidence that the students coming from these programs will have an enormous impact on the lives of millions of the poorest people in the world.

Adlai Wertman is Founding Director of the Society and Business Lab at USC Marshall School of Business.

A Transformational Partnership

By Mariana Amatullo

ENGAGING STUDENTS to experience art and design education for social impact beyond the studio's walls has been a foundational tenet of Designmatters. With a rather bold determination to bring the talent and excellence of the Art Center creative community to bear on some of the most intractable issues of our time, we like to look at the world as our classroom, with an eye toward changing it for the better.

The story of the Safe Agua project starts at the end of 2008, when Rafael Achondo, the Director of Development for the Chilean-based nonprofit organization *Un Techo para mi País* ("A Roof for my Country"), first visited our Pasadena campus with an invitation to collaborate. It was our first introduction to an organization that started as *Un Techo para Chile* (Un Techo) in 1997 by founder Felipe Berríos, a Jesuit priest who also happens to be a social entrepreneur. Berríos set about to change the face of poverty in Chile's slum communities with a sustainable and inclusive model for development that has produced nothing short of outstanding results. Addressing systemic poverty by flattening social barriers and discarding an "us versus them" view of the world, the model mobilizes the best and brightest university youth across Latin America to volunteer and meet the needs of slum residents by building transitional housing as the first stage of a comprehensive program that integrates a series of long-term skill development services focusing on empowering individuals, families, and communities, with the tools to take charge of their own future.

This is not about parachuting handouts of aid, but connecting often divided sectors of society, and ultimately building the commitment necessary for a more humane and prosperous outlook for all. Today, Un Techo's headquarters on the outskirts of Santiago employs more than 100 young professionals recruited from the top universities in the region. They lead an organization that has rallied high-level public and private sector support, and expanded to include national chapters in another 15 countries of Latin America, serving over 200 million people who live in extreme poverty throughout the continent.

Safe Agua marks the initial outcomes of the partnership we formed with Un Techo and the extraordinary team of its Social Innovation Center (*Programa Mínimo*) led by industrial designer Julián Ugarte. Without a doubt, this is an exceptional project emerging from the Designmatters network of international alliances that aligns all the ingredients for meaningful engagement: one that creates space for everyone to cross-pollinate expertise and create new value. In many ways — even before any of us boarded the plane bound to Santiago in August 2009 to meet the full team of the Innovation Center and the families of the slum community of Campamento San José where we would work — we knew we had signed on for a journey that would become much more than another impactful class. Yet we could not have anticipated how deeply transformative the personal connections we made along the way would become, nor how significant the solutions that arose would be.

Beyond the processes and solutions of a well-honed brief, this book compiles the voices and perspectives of a multidisciplinary design team of Art Center students who brought unwavering perseverance to the search for meaningful ideas that would become actionable, appropriate, and affordable designs for a community of users whom they learned to know by first name. The pages that follow present the thorough documentation of the field research and design-thinking methodologies that the team embraced in order to develop the six interrelated solutions that address the quotidian challenges of consuming, transporting, and conserving water in a setting where a biweekly truck delivery is the only source of clean water.

If this publication captures a bit of the magic of our collaboration with Un Techo, it is also an attempt to reflect upon its relevance in the context of a global knowledge economy that is redefining the role of art and design as a potent catalyst for social innovation. This broader reflection benefits from expert perspectives that we are delighted to be including: in the foreword, that of Professor Adlai Wertman, founder of the Society and Business Lab at USC's Marshall School of Business; and in the "Big Picture" section, that of Dr. Peter Gleick, cofounder and president of the Pacific Institute in Oakland. Gleick's article not only lays out an important historical overview of the global water crisis, but also offers thought-provoking questions about the leadership that our institutions will need to exert in order to safeguard this vital resource for future generations. Patrice Martin, from IDEO's Social Innovation team, provides important insight about the human-centered and open source design tool kit IDEO developed to "hear people's needs in new ways, create innovative solutions to meet these needs, and deliver solutions with financial sustainability in mind." The IDEO model of best practices was an important inspiration for the Safe Agua team's own design-based research methodologies to problem solving.

The contributions of the lead faculty team of Safe Agua — Liliana Becerra, Dan Gottlieb, and Penny Herscovitch — are at the center of this volume. Their impassioned commitment to this project is at the root of the success we celebrate now, in witnessing the positive feedback from field testing of several of the prototypes developed by the students, and the imminent implementation of these proposed designs.

Shortly after the powerful earthquake and tsunami that hit Chile's coastal communities in February 2010, Un Techo was designated the official nonprofit organization in charge of the country's emergency housing reconstruction. Suddenly, there is a new population urgently needing appropriate shelter. At this writing, Un Techo is coordinating the construction of 30,000 *media aguas* — transitional housing — all of which will go up before the South American winter months arrive. In the aftermath of this disaster, the innovative product outcomes from our collaboration, such as *Ducha Halo*, have taken on a new relevance and are being swiftly deployed by Un Techo's Social Innovation team, bringing the dignity and delight of a warm-water shower to the quake zone.

The human impact of the two weeks the Safe Agua team spent immersed with the families of the Campamento San José has transcended the pedagogical value and cultural enrichment that David Mocarski, chair of Environmental Design, and I ever expected. It is our hope that many of their testimonials, recollections, and photographs — masterfully brought to life by this publication's designers under the watchful eye of instructor Lisa Wagner — hint at the significance of the bonds that formed among all, and the trust that was established with the community. This often happened through the simple gestures of playing a game of cards, capturing snapshots with the children, lending a helping hand with a household chore, or sharing the *olla común* meal and *regatón* dance that concluded the trip. These and a collective experience of so many other heartfelt moments opened our eyes to another face of poverty, one that is not devoid of an admirable sense of optimism. Its hold in this community, as well as the importance of family, comes across in the award-winning short story *"La Ciudad de Madera,"* (The City of Wood) by 11-year-old Janitza Muñoz, a resident of a neighboring campamento.

A better tomorrow, and the promise of moving to permanent homes with running water, is also what keeps Rosita Reyes smiling. She is the inspiring community leader of Campamento San José who keeps the star-studded Chilean flag flying high on the roof of her *media agua*. "It is a symbol of hope," she says, "because this is a community with dreams."

Mariana Amatullo is Vice President, Designmatters Department, at Art Center College of Design.

Educating and Crafting Creatives

By David Mocarski

THESE DAYS, YOU HEAR THE WORDS "green," "sustainable," and "responsible" being overused, misunderstood, poorly defined, and exploited on a number of levels. As designers and educators, we must understand that what we create today could be tomorrow's tradition and history, and how we pursue the art of making could be tomorrow's craft and culture. Unless we understand the context of our projects, we are destined to make only superficial styling moves without purpose, depth, or meaning. Currently, design education has an opportunity to step up and redefine itself for coming decades. This realignment of priorities will allow us to cross-reference, reevaluate, and redefine why we design and how we approach our process. Most importantly, we have a chance to develop a stronger understanding of whom we are designing for, and why.

For us to properly inform and educate gifted young creatives, we must allow them to experience today's world firsthand. This means more than simply researching and collecting facts and data on the Web, but rather gaining real experience and making personal emotional connections. Young designers not only need to understand the depth of their creative process, but also how to integrate design, business, and culture into the world at large. They have to understand local influences, while also being able to perform globally. Today we are not educating the utopian, blue-sky student of the past, but an informed, interconnected global citizen who crafts meaningful solutions to difficult issues.

Safe Agua provided a case study in this new and necessary approach to our role in shaping the future of design, and the world. A project such as Safe Agua documents and captures the maturing of design and design education. It motivates the discipline to move past the act of designing as a linear, style-driven activity into a context-rich non-linear endeavor. Young creatives today are recognizing that before they can design, they must first consider how to interface the needs of a project, and develop a clear idea of what the emotional and physical outcomes of the project must achieve. Only then can one begin exploring actual design options.

To gain a richer understanding of this project's goals, on-the-ground field research was a necessity. Experiencing Chile and the needs of its underprivileged people firsthand presented a rich understanding of the influencers and motivators essential to developing a fruitful list of possible interventions. Meeting the people you are designing for and making a meaningful human connection establishes a bond between designer and user. And in the end, this type of social engagement helps to redefine what will be a meaningful and significant outcome. For our part, educators need to be the catalyst that brings these young creatives and global opportunities together.

Safe Agua brought together students from four Art Center majors (Environmental Design, Product Design, Graphic Design, and Transportation Design) under three instructors who also crossed disciplinary lines: Penny Herscovitch and Dan Gottlieb (Environmental Design), and Liliana Becerra (Product Design). In addressing the broad question of where our opportunities for contribution might lie, we had to ask what we, as creatives, could bring to this situation that might make a difference? How can we understand the personal issues of people who desperately need our help? How can we apply the skills we have and learn what questions we must ask in order to be contributing, making a difference, and producing design solutions that will improve people's lives? In their pursuit and their execution, projects such as Safe Agua reveal the future of design in a brilliant new light.

David Mocarski is Chair of the Environmental Design Department at Art Center College of Design.

CASCO

Un Techo para mi País/
A Roof for My Country

By Julián Ugarte

UN TECHO PARA MI PAÍS WAS FOUNDED in Chile in 1997 when a group of young college friends, supported by the Jesuit priest Father Felipe Berríos, felt the need to denounce the conditions of abject poverty in the settlements in which more than 200 million Latin Americans live. The organizers held that these conditions constitute a societal injustice for which all citizens are responsible, and must be accountable. Un Techo seeks to improve the quality of life of families living in poverty through the construction of emergency housing and the execution of social insertion plans. The work is undertaken jointly by a coalition of young college volunteers and members of the community.

Since the solidarity of young men and women knows no bounds, in 2001 the mission began spreading throughout Latin America. Currently, Un Techo operates in 15 countries: Argentina, Bolivia, Brazil, Chile, Colombia, Costa Rica, Ecuador, El Salvador, Guatemala, Mexico, Nicaragua, Paraguay, Peru, Dominican Republic, and Uruguay.

In our short life as a foundation, through the first quarter of 2009, we have built 43,000 emergency housing units. We have 200,000 young volunteers committed to transforming the bleak reality of thousands of poor families; 20,000 volunteers working in the construction sites; 5,000 permanent volunteers involved in community organizations and the social insertion programs. Our goal in Chile is to eradicate settlement camps by September 2010, by building 10,000 high-quality permanent units in order to create socially sustainable neighborhoods.

Our ultimate goal is to create a more compassionate, fair and "exclusion free" continent. One Latin America without abject poverty, with young people committed to the challenges of their own nations, where every family has a decent house and access to opportunities to improve their quality of life.

Julián Ugarte is Director of the Innovation Center at *Un Techo para Chile.*

Tools for Understanding

By Patrice Martin

I'M PLEASED TO SEE IDEO has inspired Safe Agua's iteration of the Method Cards. This small but powerful tool pushes us toward a better understanding of the world we design for — in this case, the daily struggle for access to water in the campamentos. Great design has its roots in an empathy and understanding of everyday struggles: It is not found in broad gestures, but instead the small human needs and nuances that define lives.

IDEO originally developed the Method Cards to represent the diverse ways design teams can better understand the people they are designing for. The cards provide customizable approaches and inspiration for design challenges. Using them gives all members of a team access to a multitude of problem-solving methods; even people who don't think of themselves as designers can draw out their creativity and ideas.

As Safe Agua has shown, solutions are frequently reached through details that could be overlooked. The details that matter from a design perspective can be easily deemed irrelevant or insignificant by the end user, but oftentimes are the very pieces of information that can spark a great idea. Taking a design approach draws out these small yet vital pieces of information.

For instance, in the Safe Agua project, it might not have been possible to understand the relevance of a communal laundry facility without first understanding the large role laundry plays in day-to-day campamento life. Similarly, designing a shower solution may not have seemed like a priority until experiencing the importance of dignity and pride firsthand. Relying solely on personal perspectives leads to design based on assumptions rather than real needs. Through the use of design thinking, the Safe Agua team pushed through to understand the true needs of the campamento communities.

In a similar manner, IDEO is applying its work to new challenges. For decades, it has used human-centered design to arrive at innovations for our clients — largely global corporations. But lately, IDEO is seeking to expand its reach by applying human-centered design to some of our world's most important social challenges. In this vein, IDEO created the HCD Toolkit in partnership with the Bill & Melinda Gates Foundation, and in collaboration with nonprofit groups ICRW and Heifer International. The toolkit is specially adapted for NGOs and social enterprises working with low-income communities in Africa, Asia, and Latin America. It is designed to help understand people's needs in new ways, find innovative solutions to meet these needs, and deliver solutions in a financially sustainable way.

Most importantly, the HCD Toolkit is free and available for anyone to download at www.hcdtoolkit.com. It has already been used and adapted by students, professionals, volunteers, and organizations throughout the developing world.

I'm excited to see Safe Agua building out its toolkit for social innovation. Hopefully, the cards are just one step in demonstrating Designmatters' commitment to the entire design process, where learning and understanding are essential to doing good work. Tools like the Safe Agua cards are steppingstones to a far-reaching, worldwide understanding of design for social impact. I look forward to being inspired by your next challenge.

Patrice Martin is a Social Innovation Lead and Systems Designer at IDEO.

GLOBALLY, 1.1 BILLION PEOPLE do not have access to safe, clean water for drinking and daily use. The challenge in Campamento San José is not the absolute lack of water, but rather the physical and mental burden of living without running water. It is easy for people who have running water to take it for granted. We bend water to flow through our lives — it appears at the turn of a knob and disappears down the drain. In the campamentos, people bend their lives to accommodate the realities of water.

EVEN THOUGH I LOVE MY DOG,
I PREFER MY FELLOW MAN

By Julián Ugarte

IN THE YEAR 2007, walking by a prestigious Engineering School in Chile, I witnessed how junior students, together with their professor, developed a new GPS application in their classrooms. The device was designed with nanotechnology and the shape of a leash for pets, in the case of Chile, dogs. In this way, if dogs get lost in the big city, their masters could locate them fast and bring them back home, in order not to mourn the loss of a pet that, over the years, has become another member of the family.

While we were looking around the room where these "out of the box" technological prototypes were exhibited, and their smart creators, probably the smartest and most highly trained college students of Chile, defending their proposals with remarkable conviction and determination, we could not do anything but congratulate the team that undertook this initiative. However, deep inside we had a deeper concern that took us to a scenario of uncertainties, challenging us to look even further:

How is it possible that in a country like Chile, located in a continent with 200 million poor people, the time and resources of the best college students are invested in creating a solution for pets and not for human beings?

When did colleges forget that the professionals they train are a small percentage of the population, the intellectual and financial elite, who are in charge of "driving the country forward," and helping their fellow countrymen who did not have the same opportunities? It seems that, suddenly, huge corporate buildings and development pictures scattered all over our city made us forget that in our country and in our continent there are people in need of urgent solutions, like the 80 million Latin Americans who experience what living in abject poverty means on a daily basis.

I do not know how many dogs live in Chile, but I did not want to search for that on the Internet because I sensed it was less than the 13.5% of our population living in poverty. And even though I may be wrong and there are more dogs, I consider it more important and urgent to come up with a solution for this so needy 13.5%, because even though I love my dog, I prefer my fellow man.

The answers to these difficult questions did not take long in coming: the prevailing logic of the global market shows us that solutions developed in an industry benefit the people who can afford them — it's that logical and simple. That is why the poorest families, who are not considered "valuable consumers" due to their low incomes, are alien to the market world and, thus, also to the innovation and development of new products and/or services focused on meeting their specific needs, such as surviving for example.

Therefore, paradoxically, the ones with more needs are also the ones with fewer solutions, and sadly, the market and innovation centers play this game, excluding the poorest for being poor. This is how the saying is reaffirmed: "Wealth creates more wealth." But it entails another saying: "Poverty creates more poverty."

This is the reason why in that room of young engineering students we saw sharp solutions for dogs, because their masters would make an attractive market niche to spend time and "brains" on in order to satisfy their latent needs, over the urgent needs of another group of people with less resources to pay for them. We were facing a structural problem of our society that goes beyond the limits of those engineering classes and embraces an entire continent.

Two days after the visit to that prestigious college, we had a meeting with a humble inhabitant of a campamento located "on the other side of Santiago," that Santiago de Chile unknown by many, which is far from our homes, colleges and tourist attractions.

She told us how the place where she lived was like. She said that "campamentos are irregular plots of land," that they usually receive water from a municipal truck and keep it in plastic containers outside their homes. "Once water is in the containers, we pour it in bowls to use it for our daily activities." She told us that they took partial showers and that 28,000 families in Chile still live in these settlements.

She also told us that in her community there was no sewer system, that feces were disposed of in black holes, so they lived side by side with flies, bad smells, and diseases.

Despite everything, they were happy to live there because after 10 years they had received an offer from an electricity company which was willing to sell electric energy to their community, as to any other Chilean.

After imagining the environment where she lives, and trying to interpret how she bathed in parts, we asked Cecilia, our campamento friend, how she was going to go back to "her Santiago" from the café in "ours" where we had met. The answer couldn't have been simpler — "by bus," she said.

That simple answer triggered what the Un Techo Innovation Center is today. If we think about the bus Cecilia takes, despite that she has a destination different from ours, it has the same seats, colors, drivers, and cost — about one dollar per ticket. At that point we understood two things:

If the bus is the same, probably the soap Cecilia uses is also the same (and costs the same). If Cecilia takes partial showers, i.e. she bathes differently, why is the soap the same even though it was designed for our shower and not hers (which she doesn't have)? This is the first manifesto: the market can do something more to help Cecilia, mainly because up to now it has done very little for her and her neighbors.

If they are paying the same for something, i.e., they are valuable consumers, and have an unmet need (the shower), why don't we "get our act together" and get the industry involved in developing a new type of inclusive products, like a soap for camps that works without showers and costs the same or a little less. Why don't we place families living in poverty on the innovation focus, particularly if we consider them now as valuable consumers?

Largely inspired by the work of professor C.K. Prahalad, and driven by the deep conviction of Un Techo professionals and volunteers that allows them to change the unfair and excluding reality of their countries, we decided we couldn't sit and watch how this takes place, how young professionals are being trained to serve themselves and not their country or the world while millions of Latin Americans literally live in the mud.

We saw here a wonderful opportunity for dialog and comprehensive socioeconomic and academic development.

We decided to make a technological revolution by creating the largest Social Innovation laboratory worldwide, with an army of thousands of young college brains and settlers like Cecilia, who have the capacity to transform social problems into specific and comprehensive solutions together, promoting from the collective "work," the adoption of new rules in the market and colleges.

Within this framework, the historic Alliance with Art Center was made. It is one of the world's best schools of product design, which has brilliant students and professionals, who I looked with admiration from my school when I was studying design. Together with the campamento families in Chile, we decided to embark on a trip to show the rest of the society that a change can be made, that design and innovation are lethal humanitarian weapons, which allow us to make a substantial difference between what we see today and what may exist tomorrow.

This beautiful work shows in each of its lines the commitment of the Innovation Center and Designmatters. All this has arisen from a great many 21st Century professionals and students who prefer big challenges over a big salary. Our fellow men and women, like Cecilia, cannot be absent in big ideas or challenges, especially now that we know from empirical experience in campamentos that they do not want to get things for free, but they want to have the chance to build something better for everyone together.

This shows that despite the social conditions that separate us, there are no first or second class citizens; we are all equal; we breathe, eat, and bathe; we are human beings, citizens of the world who have collectively written the history of our planet, and therefore, we can continue writing what we want for the future through our actions.

On the understanding that we are part of the world, we have the right and duty to help in writing the lines of the future we dream of. This is what we have been doing with all our hearts and minds, from the inclusive dialog to the innovation among different social sectors. This book of dreams, which I wish become part of tomorrow's history, is to a great extent entitled *Safe Agua*.

AGUA

ETHNOGRAPHIC FINDINGS
By K.C. Cho

Shortage of freshwater supply has reached a critical level. Overpopulation, economic growth, pollution, mismanagement and climate change are some of the main factors contributing to the world's water crisis.

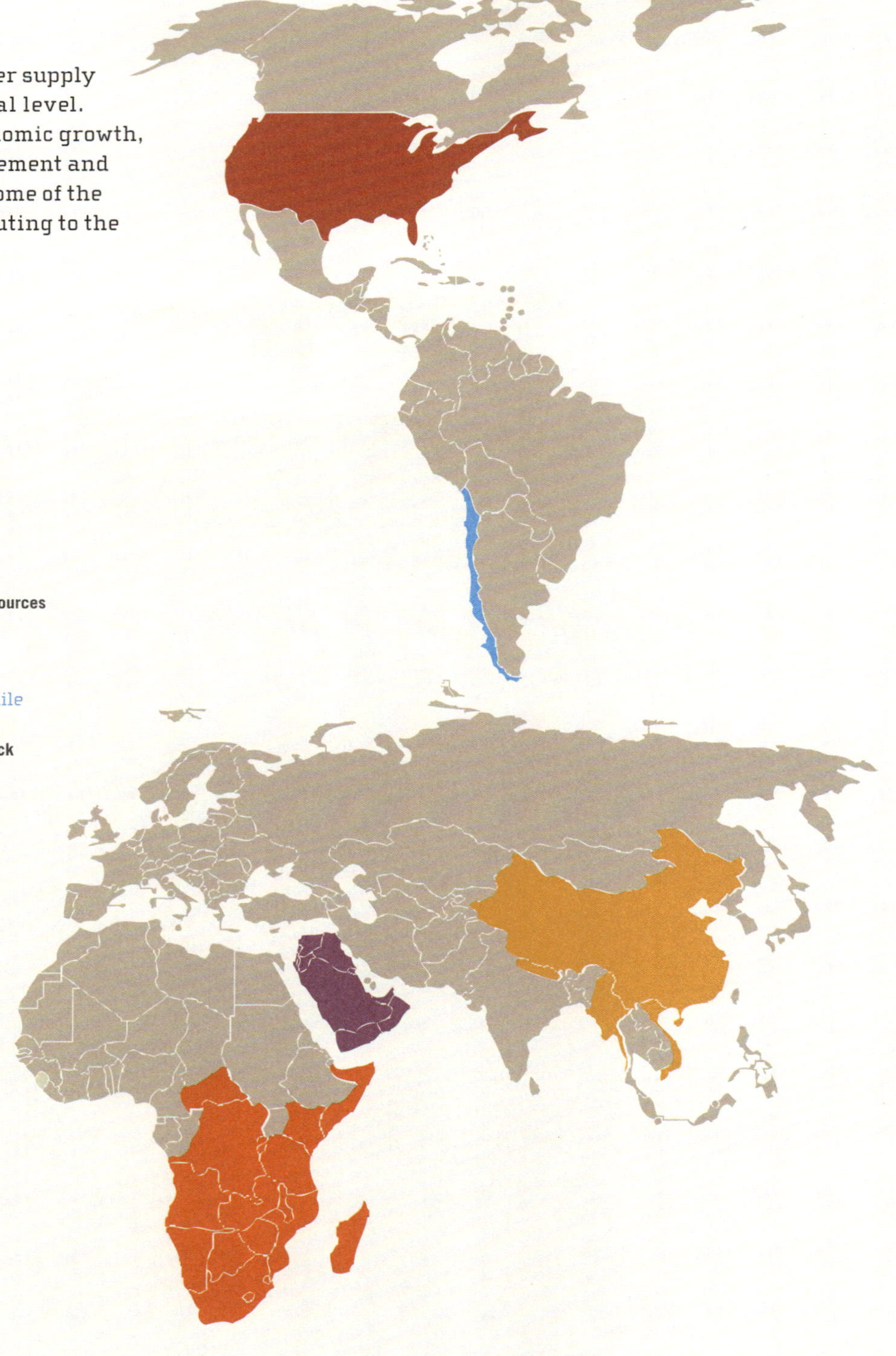

NORTH AMERICA
Southern California
- Drought
- Mandatory water restrictions
- Over-exploitation of water resources

LATIN AMERICA
Slum Village, Santiago, Chile
- Privatization
- Transportation of water by truck
- Pollution

MIDDLE EAST
Lebanon, Jordan, Israel, and Palestine
- Countries fight for control over Jordan River
- Water desalination

AFRICA
- Water scarcity
- Water-borne health issues

ASIA
Fujian, China
- Suffering crises in rural areas
- Exploding population
- Industrial waste pollution
- Rise in cancer fatalities

FACING DOWN THE HYDRO-CRISIS

By Peter H. Gleick

IN MARCH 1997, IN RURAL HEBEI AND HENAN PROVINCES IN CHINA, SEVERAL HUNDRED FARMERS FROM NEIGHBORING VILLAGES CLASHED OVER ACCESS TO WATER RESOURCES TO IRRIGATE THEIR CROPS, LEADING TO DOZENS OF INJURIES. THE RIVERS THAT USED TO SUPPLY ALL THEIR NEEDS WERE DRYING UP. TWO YEARS LATER, VIOLENT CONFLICTS OVER WATER ESCALATED IN THE SAME REGION. HUNDREDS MORE VILLAGERS WERE INJURED AND WATER DIVERSION FACILITIES DESTROYED. / IN 1999, SOME 700 SOLDIERS WERE SENT TO QUELL FIGHTING THAT CLAIMED SIX LIVES AND INJURED 60 OTHERS IN CLASHES THAT ERUPTED BETWEEN TWO YEMENI VILLAGES FIGHTING OVER A LOCAL SPRING. IN 2001, CIVIL UNREST IN PAKISTAN OVER SEVERE WATER SHORTAGES LED TO PROTESTS, RIOTS, AND BOMBINGS, KILLING ONE AND INJURING DOZENS. IN 2004, A SIMILAR DISPUTE IN A BORDERING REGION OF INDIA LED TO FOUR DEATHS AND MORE THAN 30 INJURIES. / BETWEEN 2004 AND 2006, AT LEAST 250 PEOPLE WERE KILLED AND MANY MORE INJURED IN SOMALIA AND ETHIOPIA IN FIGHTING OVER WATER WELLS AND PASTORAL LANDS. VILLAGERS THERE CALL IT THE WAR OF THE WELL AND TELL STORIES OF "WELL WARLORDS, WELL WIDOWS, AND WELL WARRIORS."

Across the globe, these sorts of public protests, disputes, and violence over water are increasingly common as problems of contamination, shortages, and allocation grow.

In rural villages and expanding cities around the world, water is an increasingly scarce and contaminated resource. As populations and water demands continue to expand, the heightened risk of violent conflicts over water use and contamination suggest new calls for fundamental changes in the way we manage and use this precious resource. The world of water is changing — not just how much water is available, or who controls it, but the whole way we think about and manage this precious commodity. The assumptions we made in the last century about the availability and use of water no longer seem to apply. And for water managers, planners, hydrologists, engineers, economists, policy makers, and concerned citizens, the time has come for new thinking and new solutions.

Over the past several centuries, societies have developed different technologies, practices, and institutions for supplying safe and reliable freshwater, dealing with extreme events, as well as collecting, treating, and disposing of wastewater. These tools brought enormous benefits to human kind. But they have also failed to solve some of our most difficult water problems, and in key ways they are unsuited to our new challenges. We need a new approach — movement into what I call the Third Water Era.

THE FIRST TWO WATER ERAS

The First Water Era lasted for some millennia, before human civilization evolved out of the most primitive hunter-gatherer existence and struggle for survival. The earliest societies relied on the natural hydrologic cycle to provide water for their use and take away what they didn't want. Put more simply, rivers and streams brought drinking water and fish, and washed away untreated detritus and human wastes. While the population of the planet was still small and dispersed, this worked well. Life was brutish and short for most people anyway, and water-related illnesses were dwarfed by the terrible consequences of childbirth, plagues, pox, and malnutrition.

Over time, this simple approach proved insufficient. As pockets of civilization began to expand and outgrow

local water resources, the Second Water Era emerged in the form of intentional manipulation of the water cycle and efforts to apply new technologies, engineering, and institutions to water problems. In the ancient cities of Rome and Greece, the agricultural fields of Mesopotamia and the Indus Valley, and other cradles of civilization, new approaches began to improve on nature's hydrologic cycle. The first dams were built across streams and rivers to divert water to farms for irrigation. The Code of Hammurabi, the judicial code dating back 4,000 years to ancient Babylonia, offers hints of the first laws and regulations governing water use, the design and operation of irrigation canals, and punishments for theft of water. Early engineers built the first canals and aqueducts to move water from places of relative abundance to places of concentrated demand. Wastewater began to be collected and isolated from day-to-day living. These kinds of innovations helped early populations live longer, interact more closely, and create cultures of art, philosophy, and science.

In some ways, however, the Second Water Era truly began in the mid-1800s, when versions of our current approaches to water management and use were developed. Cities in industrializing regions were then outgrowing and contaminating their water supplies; waves of cholera and other water-related diseases were sweeping the world; and human scientific and engineering ingenuity was blossoming. By the middle of the nineteenth century, new tools of observation, statistics, and epidemiology were being tested and there were clues that many health challenges were the result of contaminated water and bad water management.

In 1854, Dr. John Snow, a London physician, conducted a simple yet brilliant test that helped to settle the debate about the transmission of cholera. One of the poorest neighborhoods of London — served by central wells and lacking sewage collection — had been beset by a virulent cholera outbreak. He plotted the homes and numbers of people affected and noted the location of the wells that provided water for the hardest hit neighborhoods. He concluded that the source of contamination was the water from one particular well on Broad Street. He received permission from local authorities to remove the pump handle, which forced residents to draw water from other, uncontaminated wells. Within days, the outbreak subsided.

As science and medicine revealed more about the sources and prevention of water-related diseases, a revolution in thinking about water swept through the rapidly industrializing world, leading to sewage systems, innovative water treatment, new piping and distribution investments, and efforts to clean up and protect drinking water sources. This era also saw the first physical, chemical, and biological treatment systems for large, centralized volumes of waste. The first dams of gigantic scale were built to hold back floods, supply water in dry periods, and produce reliable, clean electricity. The technology was developed and deployed to build aqueducts — not tens of kilometers dug out of dirt, but thousands of kilometers in length, through or over mountains, from glaciers to the deserts. Large-scale irrigation systems were designed to permit farmers to grow food in places and at times never before possible.

What were the consequences of these advances? Cholera and dysentery, rampant in cities like New Orleans, Philadelphia, Chicago, New York, London, Paris, Moscow, and other major urban centers in the 1800s, were vanquished in developed nations, largely through the use of chlorine, filtration, and other wastewater treatment processes. The Green Revolution, due as much to the modernization of irrigation as to the application of fertilizer and pesticides, helped hundreds of millions of people avoid massive starvations in the twentieth century as the global population quadrupled from 1.5 billion to 6 billion. And nature's fury — unavoidable floods and droughts — are at least partially controlled and less damaging.

THE THIRD WATER ERA

But despite these enormous advances; despite our better understanding and technology; despite the hundreds of billions of dollars spent by utilities, towns, nations, and the international community; we still face a global water crisis of a magnitude unlike any before in human history. Not only have we not solved all of our traditional water problems, but we are now faced with new and difficult challenges. Today, water is taking center stage as the most critical resource issue facing humanity. First, there's the complicated question of supply. The total amount of water on the planet is fixed — neither growing nor shrinking. But as the population continues to grow, per-capita water availability is declining. Moreover, while this crisis is global, the impact is felt most acutely on the regional level. Water demands in some regions are rapidly increasing, as economic growth, new industries, and new technologies produce new and higher water demands. Roughly one-third of the world's population still lacks access to the most basic water services, including 1 billion people without any access to safe and affordable drinking water and 2.4 billion without access to adequate sanitation. The harsh reality is that there are hundreds of millions of cases of water-related diseases and some 2–5 million deaths per year, largely young children. Most are entirely preventable. At the same time, industrial activities are contaminating water with vast quantities of man-made pollutants — from the gasoline additive MTBE to perchlorate, endocrine disruptors to pharmaceuticals — the effects of which we only poorly understand. As a Kashmiri proverb warns, "It is easy to throw anything into the river, but difficult to take it out again."

As the planet's population surges and a new, global middle class emerges with increased appetites, especially for water-intensive meat, food production has become a critical component of this water crisis. While the area of irrigated land is growing, it is doing so at a declining rate, putting increased pressure on agricultural production. Many regions of the world are suffering from rapid depletion of groundwater resources that are being pumped faster than nature can replenish. By some estimates, as much as 40 percent of our food production comes from such unsustainable water resources.

In our oceans and rivers, a growing number of fish species are threatened or endangered by the human use of water. Some aquatic ecosystems have been completely destroyed or irreversibly modified by human water withdrawals. For example, the Aral Sea, nestled on the frontier between Kazakhstan and Uzbekistan, was once the fourth-largest inland saltwater body. Today, it is barely a quarter of its size a half century ago — thanks to the massive diversion for Soviet irrigation projects of the vast rivers that once fed it. All 24 species of fish found only in the Aral Sea are now extinct. Likewise, nearly one-third of all North American freshwater fauna populations are considered threatened with extinction, a trend mirrored elsewhere around the world. Water flows in average years no longer reach the deltas of many of the world's great rivers, including the Nile, Yellow, Amu Darya, and the Colorado, leading to nutrient depletion, loss of habitat for native fisheries, plummeting populations of birds, erosion of shorelines, and adverse effects on local communities.

All of these problems are likely to be made worse by the world's changing climate, which will have an increasing impact on water resources and the systems we built to manage them. As temperatures rise, the need for water will rise; as precipitation patterns change, water availability will change. Glaciers and snowpacks are diminishing, while the frequencies and intensities of storms are more irregular. Meanwhile, water managers are wholly unprepared to meet the demands of a different climate. The final issue, of course, is the world's broad and all but total failure to integrate questions of growth, development, and resource use, of which water may be the number one victim. These pressures come at the very moment the competition for limited freshwater is growing among users, threatening local, regional, and potentially, global stability.

NON-TRADITIONAL REACTIONS

The first reaction to our water problems is often a traditional, knee-jerk response — that we just need to pay more attention, and put more effort and more money into addressing this crisis. Doing more of the same, however, will not be enough. In fact, while the tools and methods used in the Second Water Era brought great benefits, they also brought huge and unexpected economic, social, political, and environmental costs. It is high time to look for new ideas and answers. Over the past two decades, with little fanfare and recognition, a new way of thinking about water has been taking shape. Community-scale sanitation and hygiene projects have been developed, tested, and implemented. New technologies for water treatment are being brought to market. And innovative forms of information and communication are appearing at an astounding rate, permitting successful solutions to be more quickly and widely appreciated and implemented.

ALL OF THESE PROBLEMS ARE LIKELY TO BE MADE WORSE BY THE WORLD'S CHANGING CLIMATE, WHICH WILL HAVE AN INCREASING IMPACT ON WATER RESOURCES AND THE SYSTEMS WE BUILT TO MANAGE THEM.

This new era requires nothing short of a revolution in thinking about water — a fundamental re-evaluation of water planning, policy, and management. We need to utilize technology, environmental science, economics, and new institutional approaches to address unresolved water challenges and to tackle new threats, such as climate change. We need what the Pacific Institute calls a Soft Path for Water.

First, the top priority is to meet the basic water requirements for all humans and all ecosystems. Governments, international aid organizations, corporations, and private groups at all levels must join forces to meet the goal of providing these basic needs universally. The economic costs of meeting these needs are far less than the economic costs of failing to do so. Some efforts are currently underway. But funding, agreements to transfer technology and information, and institutional efforts on the part of international agencies and local governments are all inadequate. For basic human needs, the efforts of the world community through the Millennium Development Goals are a key beginning. These goals include the targets of expanding access to water and sanitation for the world's poor. For basic ecosystem needs, there are new efforts to identify, define, and satisfy water requirements through policy initiatives. In North America, efforts to provide basic river flows or restore ecosystem health are underway in the Florida Everglades, the Great Lakes, the Sacramento/San Joaquin Delta, and many other places. In South Africa, the constitution guarantees meeting the basic water needs of both humans and the environment. And in Europe and Asia, there are proposals for new standards for environmental management. Just as in battling climate change, the United States has an opportunity to be a player in solving water issues through smarter funding and the utilization of our vast scientific, technological, and educational resources. But we are not yet playing that role to the extent of our ability, either at home or abroad.

Second, we must rethink our approach to both water supply and demand. On the demand side, there is vast potential to become more efficient in how we use water, in every community, in every use — from industry to commerce, from homes to farms. Our societal goals are not the "use" of water, but providing goods and services to society: getting rid of wastes, producing industrial output, washing our clothes and bodies, and growing more food. While most of these steps require water, there is tremendous potential to reduce the amount of water needed to accomplish these goals. If we can do these things with less and less water, we will be better off. We are already making progress in this area: overall water use in the United States is lower today than it was some 35 years ago; China is beginning to raise water prices to encourage conservation and efficiency; Mexico City is beginning a major effort to find and stop leaks that prevent water from reaching consumers; and Singapore's water conservation programs, coupled with innovative new supply developments, have helped reduce its dependence on imported water from Malaysia.

It is possible to break the assumed connection between economic growth and water use. Indeed, improving efficiency is perhaps the most important new tool in our arsenal.

At the same time, we must also rethink our water supply alternatives. The traditional approach of building dams, aqueducts, and central irrigation systems to take more water from vulnerable ecosystems and watersheds is unsustainable. New approaches must be developed. In many countries, new water management infrastructure is still needed, but it will have to be built to better economic, social, and environmental standards. The days of misleading and incomplete cost-benefit analyses — ecosystems ignored, social impacts hidden, and communities left out of the decision-making process — must be brought to an end.

But we should also expand the definition of what we consider new "supply." Water supply must also include increased use of recycled and reclaimed water, rainwater harvesting, desalination (where appropriate), and innovative blended use of surface and groundwater. In Namibia, Singapore, and California, wastewater is increasingly treated to a very high standard and reused for industrial or commercial purposes, even for drinking. And in India, traditional rainwater harvesting is once again restoring groundwater and stream flows, reviving communities and giving new options to struggling farmers.

WATER SUPPLY MUST ALSO INCLUDE INCREASED USE OF RECYCLED AND RECLAIMED WATER, RAINWATER HARVESTING, DESALINATION (WHERE APPROPRIATE), AND INNOVATIVE BLENDED USE OF SURFACE AND GROUNDWATER.

Third, we must do a better job of protecting water quality and matching requirements to demands. Why is expensive potable water still widely used to make our golf courses green or flush our toilets, when other water is readily available? In the western United States and other regions, where water is scarce, lower quality alternatives are already being used. In developing nations, new technologies are beginning to offer better, cheaper, and more reliable water-quality monitoring to protect human health. But governments must improve and strengthen standards for water quality and more aggressively enforce those standards. Greater attention to water quality will not only save lives and protect ecosystem health, it will also expand the amount of water resources available for use. In China, for example, as much as 40 percent of water is too contaminated to use for domestic and agricultural purposes. Accelerating the construction of modern waste-water collection and treatment systems and enforcing laws against polluters can help stop the needless sickness and death that result from the consumption of dirty water, while simultaneously increasing the volume of water that can be used to meet agricultural requirements — and that now simply goes to waste.

WHAT WILL THE FUTURE BRING? ARE WE AS HUMANS CONDEMNED TO A PERPETUAL WATER CRISIS AND GROWING CONFLICTS OVER WATER?

Fourth, global climate changes are coming; indeed, they are already upon us. Humans are altering the composition of the atmosphere in a way that is guaranteed to alter the climate for centuries to come. The scientific evidence that climate change is already affecting our water resources is rapidly accumulating. We are seeing the loss of snowpack in the world's mountain ranges. The glaciers of the Rockies, Himalayas, and Andes are fast disappearing, which will affect downstream communities that depend on the water stored in these mountains. The timing of rainfall and runoff is shifting. Billions of people and the coastal infrastructures on which they depend are vulnerable to storms and rising sea levels. These risks impel legislation to reduce greenhouse gas emissions and plans for adapting and mitigating the impacts that are now, sadly, unavoidable. With regard to building the facilities required for sustainable water management, climate change must be considered. Wastewater treatment plants and desalination facilities along coastlines must take into account sea levels of the future, not of yesterday. And ongoing projects to dam rivers and tap into ground-water aquifers must consider how future climatic condi-

tions will affect operations and hydrology. In short, we must manage our water systems for tomorrow's climate or face the reality that they might be obsolete before the final brick is laid.

Finally, we must improve institutions that manage our water resources, including utilities, planning agencies, government bureaucracies, and companies. Charles Darwin observed that "if the misery of our poor be caused not by the laws of nature, but by our institutions, great is our sin." Well, we must then repent. The failure to solve our water problems is not the result of a lack of water, money, or knowledge. Ultimately, it is the failure of our institutions. Universities will have to broaden what they teach about water — expanding beyond simple engineering to smart economics, sociology, and political science. Water utilities must devise innovative environmental management systems that permit them to sustainably satisfy the needs of both human and natural ecosystems. Government planning agencies must make real efforts to integrate growth, planning, and water resource issues by prohibiting development in regions with inadequate water resources or limiting the kinds of development to "green," low-water-use design and construction. New collaborations are required among farmers, environmentalists, industry, and water agencies to reduce the risks of violence over diminished water resources. And an innovative set of economic tools must begin to set reasonable pricing incentives to encourage efficient and sustainable use of water and eliminate subsidies that promote bad practices.

What will the future bring? Are we as humans condemned to a perpetual water crisis and growing conflicts over water? This unwanted and dangerous path can now be seen far more clearly than ever before, and at this critical juncture, every step we make will affect the course we follow. As population growth and development pushes us even closer to the limits of our water supply, our focus must shift to a new way of thinking. Water is a precious, scarce, and vital resource and our use of it must be thoughtful, sustainable, and carefully planned.

Peter H. Gleick is Co-founder and President of the Pacific Institute in Oakland, California, a MacArthur fellow, and member of the U.S. National Academy of Sciences. He is the author of seven books, including the biennial report, *The World's Water* (Island Press).

By Giancarlo Llacar and Evangeline Joo

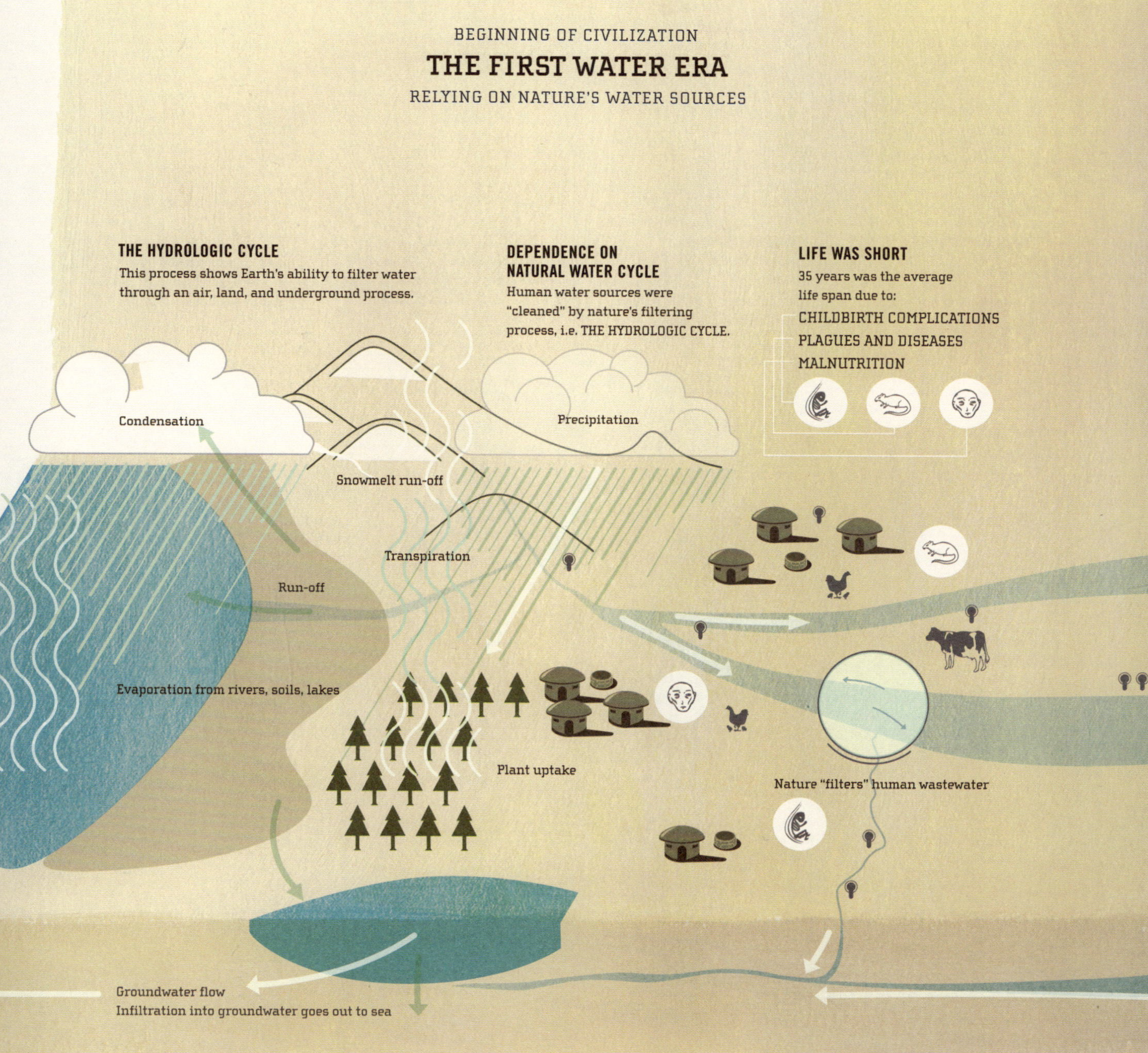

START OF 1800s

THE SECOND WATER ERA
MANIPULATING WATER SOURCES

HUMANS MANIPULATE WATER CYCLE
A higher standard of living is achieved through innovations in water technology such as DAMS, IRRIGATION, and CANALS.

WASTEWATER IS COLLECTED & ISOLATED
Canals help strain dirty water out of main tanks, thus fewer deaths occur in the big cities.

POPULATION WAS LIVING LONGER
With less water-related diseases affecting communities, fewer people were dying.

OVERDEVELOPMENT
New industries, economic growth, and new technology add to the demand of water worldwide.

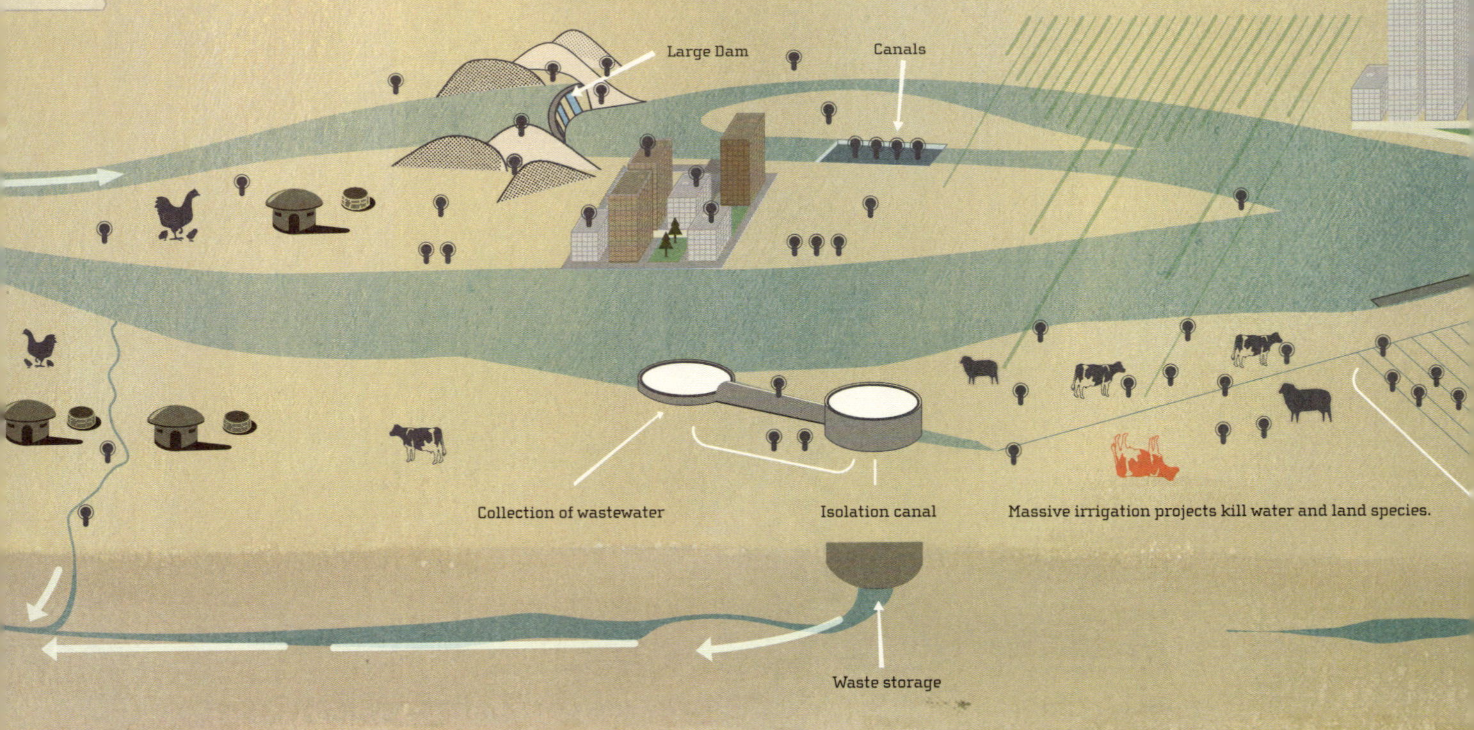

Between 2004 and 2006, at least 250 people were killed and many more injured in Somalia and Ethiopia in fighting over water wells and pastoral lands.

The world of water is changing — not just how much water is available, or who controls it, but the whole way we think about and manage this precious commodity.

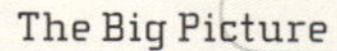

1950s–PRESENT

THE THIRD WATER ERA

MASSIVE GLOBAL CRISIS

FAILURE TO INTEGRATE
Competition for water use will
increase if the integration of
development and resource use
is not implemented.

MAJOR PROBLEM: LACK OF ACCESS
One-third of the world's population
currently lacks access to safe water,
resulting in 2.5 MILLION DEATHS
EACH YEAR.

**EARTH'S CLIMATE
IS CHANGING**
Temperatures are rising, and
the need for water will, too.
THE PRECIPITATION CYCLE
WILL CHANGE AS A RESULT.

+2 degrees in temperature
is enough to change the
ecology of the world.

Improved irrigation systems

Groundwater is depleting every year.

Nations are fighting
over water resources.

One major problem is consumption of food:
40% of all food production requires water from
sources that are unsustainable.

A municipal irrigation truck (the same one that waters Santiago's street trees) delivers water one to three times a week to Campamento San José.

FROM GLOBAL TO LOCAL:
WATER IN CAMPAMENTO SAN JOSÉ

By Dan Gottlieb and Penny Herscovitch

Families in Campamento San José receive water from a municipal truck one to three times per week. They live with the uncertainty of whether or not the water truck will arrive. When the water is delivered, they store it in barrels outside their homes. Without running water, women must hand carry water for each daily task. Bathing becomes an arduous chore rather than a relief; laundry can take a full day of physical labor; and a glass of water can make a child sick. These perpetual burdens consume people's time, diminish their quality of life, impact health and dignity, and become an obstacle to earning a stable income and overcoming poverty.

The water truck driver and his assistant fill
up each family's barrels via long hoses.

Families recieve as much water as they can store, so they
pour out any remaining old water onto the dirt. Families
store water in large 55-gallon blue barrels, 5-gallon buckets,
water jugs, other containers. Normally, a 55-gallon barrel
lasts a family for two days.

Everyday activities like drinking water, doing dishes, or taking a shower can become a burden and can even lead to injury and illness.

Families in the campamento are adept at conserving water and reusing grey water for latrines, keeping dust down, and killing insects beneath the *media agua*.

"When the water truck comes, there is happiness," said Rosita Reyes.

Dan Gottlieb and Penny Herscovitch are Faculty,
Environmental Design, Art Center College of Design.

THE PAST, PRESENT AND FUTURE OF WATER FOR FAMILIES IN CAMPAMENTO SAN JOSÉ

By Stella Hernandez

BEFORE THE WATER TRUCK

While we were walking around to get to know the campamento's surroundings better, I had a great opportunity to talk to Rosita Reyes and learn how they ended up here. This was the only time that we walked outside the campamento, having been warned not to go alone because it was not considered safe. From the campamento, which is located between a new housing project and one of the major freeways coming from the city, we could see a hill where other informal houses have been built.

We walked to the end of the campamento, and then continued a little longer until we approached an abandoned bridge over the freeway. "This was the place where they wanted us to live," said Rosita.

Years before, Rosita and her family had lived in a very informal campamento. There was no water or electricity, and conditions were very unsafe. "We used to beg for water," she said, "in order to be able to shower and send the kids to school in the morning."

I asked if anybody ever denied them water, and she said that it happened many times. "It is very hard to be refused water," she said. Her words were moving, and hard to forget.

Later, Rosita organized 20 families and started a negotiation process in order to get a *media agua* for every family; a place they could all move together. At the same time that the government was considering this area as a possible site to relocate the families, there were new housing projects being built nearby. There was some political pressure not to allow a campamento next to these new houses. It was a struggle of power and money, and Rosita and her families were in the middle, being discriminated against. That's when they were offered the option of moving onto the bridge. Rosita and all the families refused, arguing they needed a safe place for their kids to live, not an abandoned bridge over a freeway.

They went to many meetings, without results, and the possibilities of finding a new location seemed less and less likely. Then one day, much to her surprise, they were told that a location had been approved, and she was asked how many weeks it would take them all to organize themselves and move. "Three days," Rosita said, but the officials were skeptical. "I told them, you don't know what people are able to do after suffering so much humiliation. If they want a house, they will do it."

Rosita remembers that weekend very well. It was raining hard, and they did not have much to cover themselves, but they did it. In three days. Rosita's work and determination had paid off. They have electricity, the campamento is enclosed to make it safer for their kids, and water is delivered every week. They don't have to beg for water anymore. "When the water truck comes, there is happiness," said Rosita, and I certainly understood why.

I WILL BUILD A ROCKETSHIP TO THE MOON: HOW CAMPAMENTO SAN JOSÉ CAME TO BE

When walking around the campamento with Rosita Reyes, she told me her story about the rocket ship, which became my favorite story. I think it is a story of courage and kindness that shows the spirit of somebody who can work hard to make dreams a reality for the people she cares about.

There were many hurdles to overcome in Rosita's negotiations with the local government to secure a piece of land for her 20 families. There were a variety of stakeholders involved in the process, and much opposition from those involved in a new housing development being built in the area who didn't want the families to move close by. Yet Rosita persisted. She went to many meetings, patiently explaining to everybody what her requests were and why they needed the place so badly. She built a model and made a plan herself, specifying they needed 20 houses, one per family, along with electricity and water services.

In one of the final meetings, she got well prepared and sat listening carefully to everybody. Since they couldn't agree to give her the location, they made a new offer. "Do you need money?" they asked. "How much do you want to go to another place?" As upsetting as this offer was, she tried to stay calm.

Personalized hood ornament in Campamento San José.

"What I am asking for," she replied, "is not money. If you want still to give me money, I will take it and build a rocket ship. The 20 families and I will go to live on the moon, and everybody here can count on this: when you run out of resources here on earth, I will make sure that you get a piece of land next to us up there on the moon." She then stood up and left the meeting.

Rosita is a brave woman who is always standing up for the people in the campamento. She is an innate leader who has dedicated herself and worked very hard to give everyone a better place to live. The courage she demonstrated at that meeting showed what it meant to wish for something with all your heart and not to give up. Her efforts paid off, and after that meeting she was notified of the approval to build the campamento. This was the first step. She has since taken on many other challenges, working toward the biggest reward of all: 20 families will soon have a permanent place to call home.have a permanent place to call home.

PERMANENT HOUSES, PERMANENT WATER?

Casas definitivas are the projects built by *Un Techo para mi País* that provide the opportunity for people living in media aguas to buy a permanent house. Rosita Reyes and the people in the campamento have been working together, getting organized, and saving the money to fulfill the dream of getting 20 of these houses. We had the opportunity to visit two of these projects, and talked to people who have already moved into their new homes.

Apart from owning the house and not having to worry about getting evicted, they get two essential services: electricity and water. Even though these are the two services that everybody wishes for, there is a harsh reality involved — they have to pay for them, usually for the first time. Although some people have been paying for electricity already, in most cases connections to electricity and water in the campamentos are done unofficially, and are not paid for.

Right next to one of the permanent housing projects that we visited was a very informal campamento. We had the opportunity to walk through and we were welcomed to see one of the houses. After talking with the owner, we were very surprised to learn that she actually owns one of the permanent houses in the project across the street. She smiled and showed us the keys to her new house, and said that although she had owned the place for some time, she has yet to move in. When asked why she hadn't moved, she talked about how difficult is to get used to new neighbors — and also to leave the place that has been her home, where family is close by.

One of the things that we noticed in her house was that she had electricity and running water. Both services are obtained surreptitiously, but I wondered if having running water makes such a difference that she doesn't feel the urgency of moving into a new house.

Still, thinking about this case made many of us realize that owning a permanent house involves a new set of issues and challenges — emotional, financial, and social. There are difficult feelings involved in the process of leaving the campamentos and going to live with new neighbors who are coming from other areas, as well as being able to afford the expense of electricity and water services. Many people who have already moved must then start learning how to measure their consumption habits, and many people later confront the issue of not keeping water service because they are not able to pay for it.

Looking at all these issues, we realized that our research needed to go farther than overcoming the lack of safe water in the campamentos. We also needed to consider the future that people will confront when they get a new house. There should be a progression between living in *media aguas* and living in permanent houses, and water has a primary role in this transition. It is the qualitative difference between the uncertainty of obtaining water every day versus the appreciation and responsibility of having it.

New permanent social housing (*Casas Definitivas*) built by *Un Techo para Chile*.

LOOKING AT ALL THESE ISSUES, WE REALIZED THAT OUR RESEARCH NEEDED TO GO FARTHER THAN OVERCOMING THE LACK OF SAFE WATER IN THE CAMPAMENTOS. WE ALSO NEEDED TO CONSIDER THE FUTURE THAT PEOPLE WILL CONFRONT WHEN THEY GET A NEW HOUSE.

SAFE AGUA IS A UNIQUE COMBINATION of design education, design research, and social entrepreneurship. Projects such as Safe Agua are changing design education, and changing the design process itself to integrate field research as the driving component. Beyond responding to a preconceived design brief, students now are learning how to identify design opportunities and evaluate their largest potential impacts.

Our team of teachers, students, and nonprofit partners combined multiple professional backgrounds and design disciplines. This new model deeply connects people across cultures and forges alliances across borders.

Our design challenge began by asking the question: How can we work with impoverished communities (campamentos) in Santiago to develop new tools for using, storing, and transporting water in order to help families overcome the conditions of poverty?

HOW THE SATE AGUA STUDIO WORKED:
PREP IN LA
EMPANADA KICK-OFF!
SANTIAGO
STUDIO CLASS
EMPATHY EXERCISE
A DAY WITHOUT TAPS!
2 WEEKS IN CHILE
FIELD RESEARCH
14 WEEKS
ART CENTER
RESEARCH CARDS
5 MAJORS
4 TEAMS
3 SPANISH SPEAKING STUDENTS
3 INSTRUCTORS
RESEARCH INSIGHT/ OPPORTUNITY/ IDEAS
WEEK 2
• PRESENT
• RESEARCH
• INSIGHT
CAMPAMENTO SAN JOSÉ
LIVE AT INFOCAP
CASITA
WEEK 3
6 NEW TEAMS!
SKYPE WITH TECHO!
VISIT SOCIAL HOUSING
OTHER UNIVERSITIES DIALOGUE
BARREL FACTORY TOUR

WEEKS 4-5
- DESIGN STORM
- I.D. OPPORTUNITY & IDEATE
- DR. M. CASAL

WEEKS 6-7
- BIZ STORM (WHAT IS THE PROBLEM?)
- MOCK UPS FROM EXISTING PARTS

MIDTERM WEEK 8
- "FRANKENSTEIN" MODELS, NO PEDESTALS.
- SHOW PROJECTS IN CONTEXT.

- SKYPE WITH CAMPAMENTOS
- TECHO PARTNERS VISIT LA & LECTURE
- SPECIAL GUESTS!

WEEKS 9—10
- FOCUS GROUP IN CAMPAMENTO
- PROJECT DEVELOPMENT BASED ON FEEDBACK

WEEKS 11-13
- PROTOTYPE FABRICATION, CREATE INSTRUCTION MANUALS, RENDERINGS, ETC.

WEEK 14 — FINAL
- CRIT ON THE BRIDGE 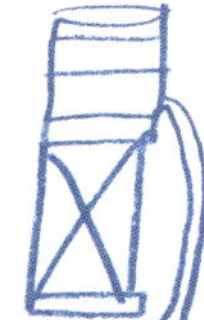
- PRESENT TO TECHO PARTNERS, CHILEAN DELEGATES, ACCD PRESIDENT & GUESTS
- WORKING PRODUCT PROTOTYPES & ENVIRONMENTS & CAMPAIGNS.

BEYOND THE TERM

SPRING 2010 — EXHIBIT IN GALLERY

TELLING THE STORY

- PUBLICATION
- DOCUMENTARY
- MOTION GRAPHICS

IN CHILE
- 20 FAMILIES TEST DUCHA HALO
- 8.8 EARTHQUAKE FEB. 27 2010
- TECHO BUILDS 30,000 HOMES FOR EARTHQUAKE SURVIVORS
- PHASE 2 PROTOTYPES

FALL 2010
- SHANGHAI, CHINA EXHIBIT/PRESENT AT WORLD EXPO
- SOCIAL INNOVATION CONFERENCE/ DESIGN BIENNIAL

"A CALL TO ACTION"

Ramon Coronado (left) and Julián Ugarte display brainstorming session ideas at Un Techo's Santiago headquarters.

A tape reminder to break the habit of turning on the tap.

DESIGN CHALLENGE

By Liliana Becerra, Dan Gottlieb
and Penny Herscovitch

FIELD RESEARCH

In order to understand and gain insight into another culture and ultimately identify design opportunities to help families overcome the conditions of poverty, we established different strategies and methodologies to gain empathy and to guide the students throughout the research phase.

EXERCISE IN EMPATHY: A DAY WITHOUT TAPS

We believe that at the root of all design is empathy. Therefore, one of our goals to begin the research process was to open our minds and develop empathy by seeking to understand people whose lives differ in many ways from our own. Establishing personal connections between students and families shifted our process from designing *for* people to designing *with* people.

For many of us, this was the first time we visited families of lower socioeconomic status, and likewise our first experience living without running water. One of the things we take for granted is convenient, unlimited water from plumbing and faucets, yet the communities we worked with in the slums of Santiago can only have water delivered once a week by truck.

That is why we decided to conduct an empathy exercise called "A Day Without Taps." Our team of students and instructors in California as well as our partners from Un Techo in Chile participated in this exercise, which helped us bond as a group and set the tone for a truly collaborative project.

For our "Day Without Taps," each student and instructor lived for a day using only a five-gallon bucket of water in order to experience the challenges that the families living in the slums face on a regular basis. We commit-

ted ourselves to use all our water for our daily activities (bathe, brush our teeth, cook, wash, drink, flush toilet, etc.) either from our nearest hose or from a previously filled five-gallon (19-liter) container. According to the United Nations, this is the average amount of water that a family in Africa consumes each day.

We each kept a detailed visual journal of our "Day without Taps," documenting with photos, sketches, reflections, and questions. We noted how many liters of water we used for each activity, and whenever possible, we used our water bill to compare how much we use on average. We often found ourselves changing our behaviors to cope with the challenge: skipping showers, postponing laundry, and coming up with different solutions to carry, store, and filter water.

Once we arrived in the slums of Santiago and started to conduct our field research, we realized that as useful as this exercise in empathy was, our experience of a "Day without Taps" was nothing compared to what people in the campamentos had to live with — not only for one day, but every day.

WATER DIARY:
ELIZABETH BAYNE
8-15-2009

TOTAL WATER USE:
12 LITERS

9 AM

2:50 PM

WASHED HANDS (.24L)
WASHED DISHES (1.89L)
WASHED FACE (.24L)
BRUSHED TEETH (.24L)

FLUSHED (4L)
WASHED HANDS (.24L)

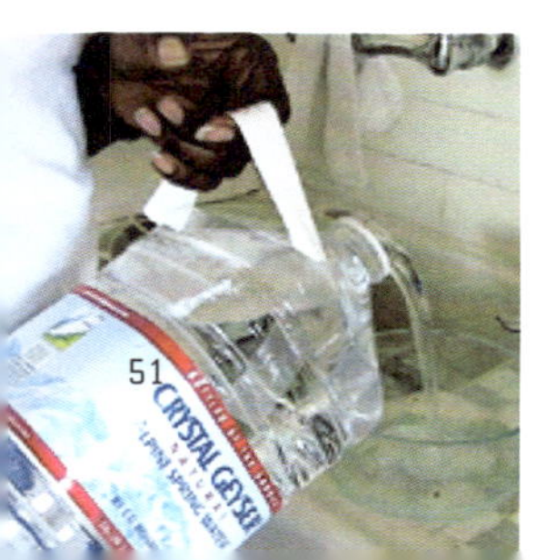

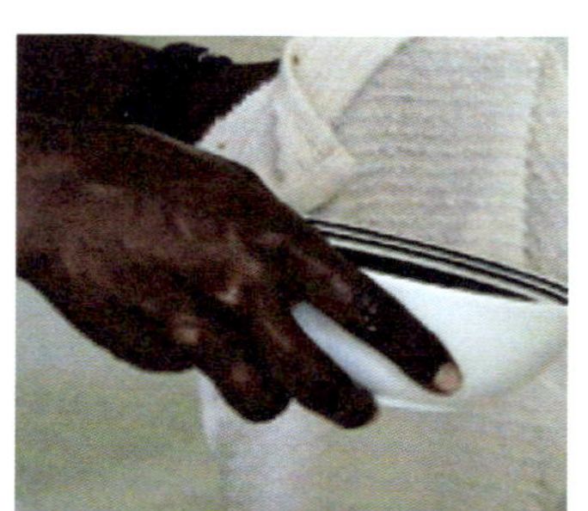

9 PM
9 AM
FLUSHED (4L)
WASHED HANDS (.24L)
WASHED FACE (.24L)
BRUSHED TEETH (.24L)
WASHED
FACE (.24L)

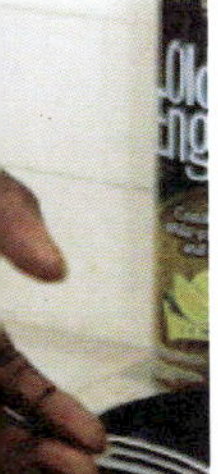
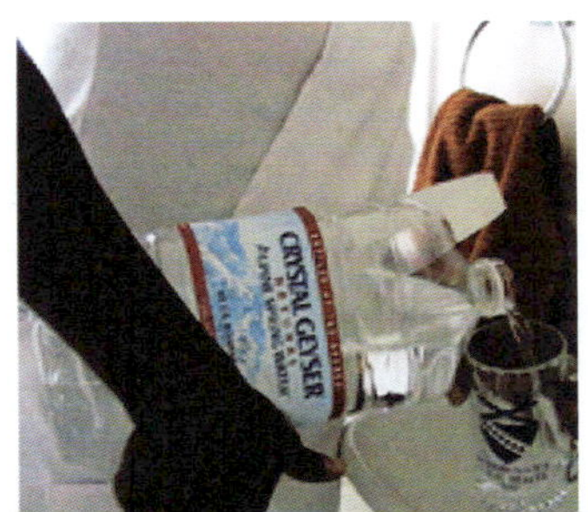
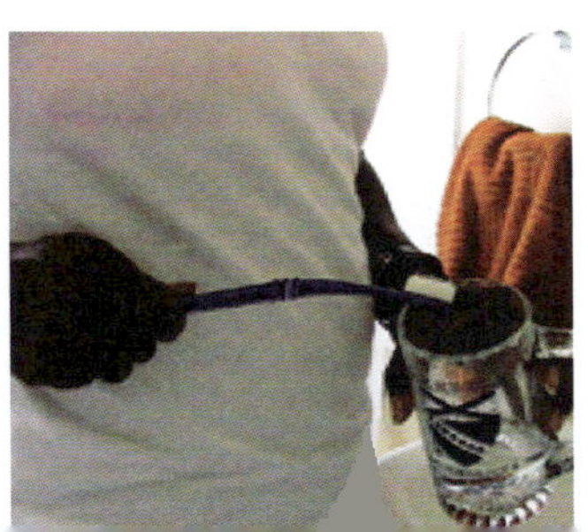
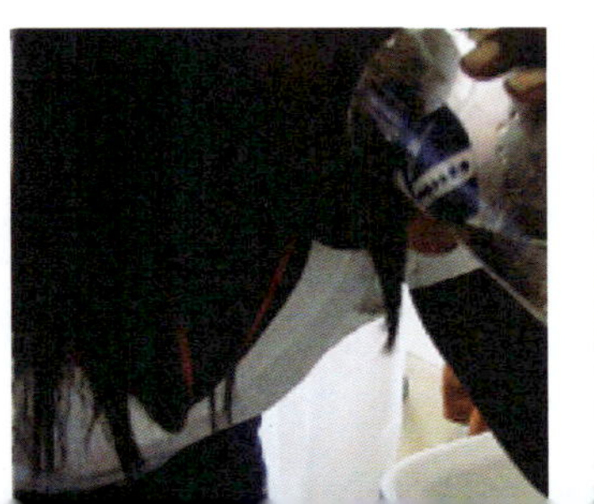
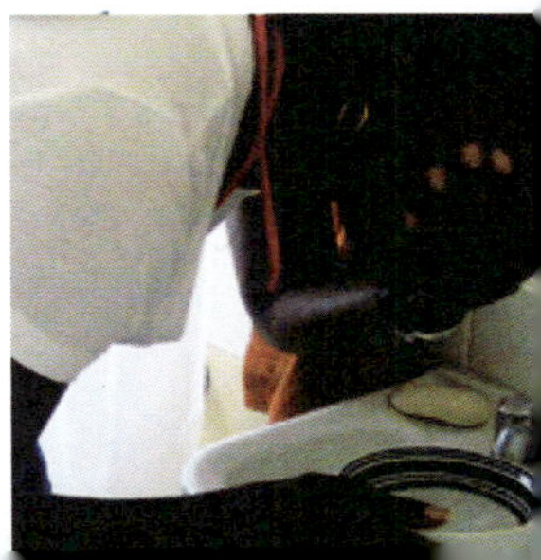

A set of methodology cards.

FIELD RESEARCH

METHODOLOGY CARDS

In order to prepare ourselves for the field research itself, we created a tool kit of methodology cards specifically targeting our project objectives. We drew input and inspiration from different design research sources and methodologies, including IDEO's method cards and their Human Centered Design (HCD) tool kit, and also from our own professional background and experience in the field of design research and insights.

The tool kit was fundamental for directing the focus of the field research. It provided our students with confidence and structure to navigate a completely new territory. It also changed the traditional model of design education by introducing field research as a key component of the design process.

The set of six cards defined the outline of the research. Each card featured one research topic and posed its fundamental questions with an inspiring image on the front and our recommended tips and strategies for gaining the relevant insights on the back. The cards were pocket size, with waterproof surfaces, to enable students to carry them out in the field as a guideline. We included a blank section on the back of the cards corresponding to the date, so students could decide and mark the order in which they would organize their research.

BROAD RESEARCH TOPICS

Rather than focusing only on the functional problems of storing, carrying, using, and re-using water, we also decided to assess the "big picture" aspects of the problem. Understanding things such as people's core values, aspirations, physical environments, and daily life gave us further insight into the problem and ultimately engaged our team on a personal level with our partners in the slums. The three broad research areas we considered were:

- ASPIRATIONS AND LIMITATIONS
- MATERIALITY AND SPACES
- A DAY IN THE LIFE

FOCUSED RESEARCH TOPICS

To focus on more specific functional, water-related issues that would directly target our project objectives and deliverables, we created the following three research topics:

- STORING / CONTAINING
- CARRYING AND MOVING
- USE AND RE-USE OF WATER

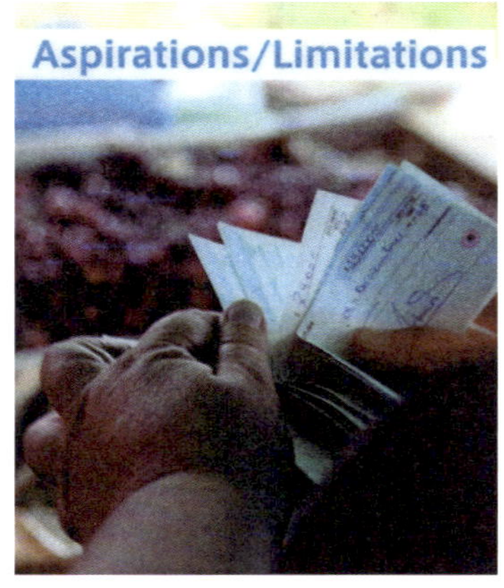

ASPIRATIONS AND LIMITATIONS

What are people's aspirations, and what keeps people from achieving them? How can we best impact this area?

PERSONAL INVENTORY (EMOTIONAL)

Document the things that people identify as important to them as a way of cataloging evidence of their lifestyles. (30 min)

COLLAGE OR CARD SORTING

Ask participants to build a collage from a provided collection of images and to explain the meaning of the images and arrangements they choose. (30-45 min)

DRAW YOUR PAST / FUTURE

Ask participants to "draw the future you want." ("If you won the lottery...") Draw a path from past to now to that future with the steps and hurdles along the way. (30 min)

EXTRA TIPS

- First, gain people's trust; gather direct, unfitered quotes.

- Plan deep interview questions and practice interview techniques; ask "why?" 5 times — to get to the real why.

- Prep and print visual cards ahead of time.

METHODOLOGY CARD 2

MATERIALITY AND SPACE
What is the material reality of personal and collective objects in the household and neighborhood? How can we best impact this area?

BEHAVIOR ARCHAEOLOGY
Look for evidence of people's acivities, habits, and values inherent in the placement, patterns, and organization of things.

SOCIAL NETWORKS & SPACES
Notice different kinds of social relationships within a user group and map the network of their interactions. In what ways do objects, materials, and spaces express social relationships?

PERSONAL INVENTORY (FUNCTIONAL)
Ask people to show and describe objects they handle daily — catalog evidence of lifestyle. (30 min)

EXTRA TIPS

- How do things wear out?

- What can we learn from resourcefulness of the material culture?

- Be aware of materials and spaces throughout Santiago, not only in the slums.

METHODOLOGY CARD 3

A DAY IN THE LIFE
Catalog a day in the life of people in the campamentos, with special attention to the role water plays. How can we best impact this area?

A DAY IN THE LIFE OF A FAMILY
Catalog the activities and contexts that water users experience throughout a day.

SHADOWING
Tag along with people to observe and understand their day-to-day routines, interactions, and contexts. (1-2 hrs)

TIMELINE
Create a branching timeline of household members' activities. Every person in the household plays a different role. How do the roles of different people relate to each other?

EXTRA TIPS

• Each team member can shadow a different household member.

• Ask the family member to record what they're doing each time a watch timer / beeper goes off.

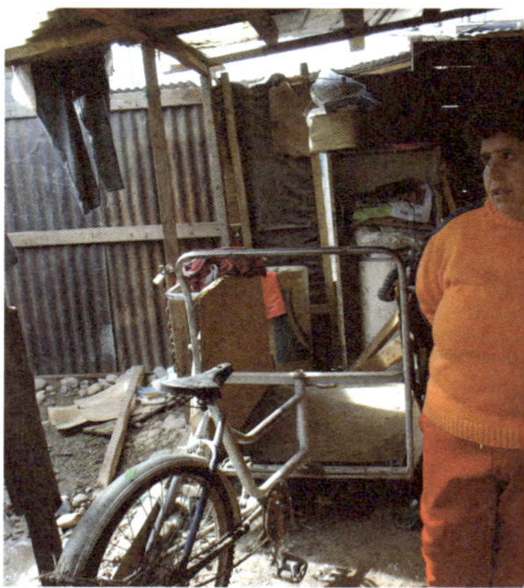

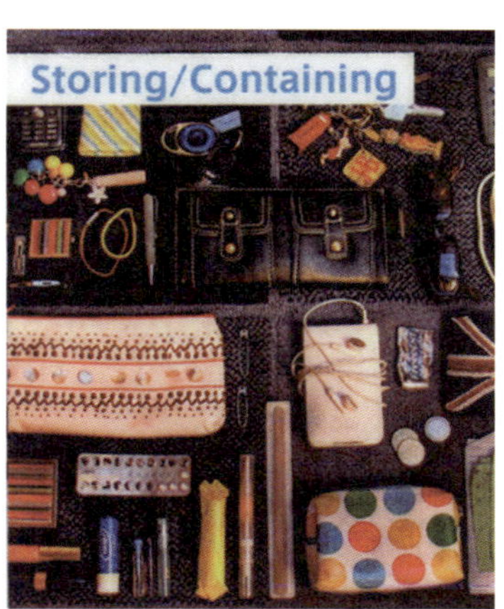

METHODOLOGY CARD 4

STORING AND CONTAINING
How do people store, contain, and protect valuables, food, water, and everyday objects? How can we best impact this area?

ERROR ANALYSIS
List all the things that can go wrong when storing/containing water and determine the various possible causes. (30 min)

SCENARIO TESTING / "WHAT IF"
After your initial research, show users a series of cards depicting possible future scenarios for storing water and invite them to share their reactions. (30 min)

GUIDED TOUR
Ask participants if you can accompany them on a guided tour of how they contain objects. Why did they choose a specific means of storage? (45 min)

EXTRA TIPS

• Be aware of cultural biases and preconceptions.

• Consider differences between storing valuables vs. daily objects.

• How does the house itself serve as a container to keep out rain, store water, etc.?

• Survey containment solutions that exist on the market and that families have invented.

CARRYING AND MOVING
How do people carry objects, water, and themselves around? How can we best impact this area?

BEHAVIOR MAPPING
Track the positions and movements of people within a space over time and note what are they carrying or moving around while doing it. (45 min)

FLOW ANALYSIS
Represent the flow of water through all phases of use. Consider water's behavior, not only on a map or plan, but also as it moves up and down.

FLY ON THE WALL
In public spaces, such as markets, neigh-borhoods, or public transit, observe and record behavior within its context, without interfering with people's activities. (1-2 hrs)

EXTRA TIPS

• What do people carry around? (Wallet, phone, children, jewelry, etc.)

• Why do they carry those things around? (Take "what's in my bag" photo)

• Survey carrying solutions that exist on the market and that families have invented.

METHODOLOGY CARD 6

USE AND RE-USE WATER

How is water used over the course of a day and week? How can we best impact this area?

STORYBOARD OF WATER'S DAY / WEEK
Illustrate a character-rich story line describing the context of use of water. Water is the main character; if water could tell its story, what would it say?

CAMERA JOURNAL
Distribute a kit with camera, journal, and instructions. Ask participants to keep a diary of activities related to using water. (15 min / 1–2 days)

NARRATION
As they perform a task or process, ask participants to describe aloud what they are thinking-to reach users' perceptions, concerns, and motivations. (45 min)

EXTRA TIPS

- Other documentation methods: Script photos — ask people to re-enact each step of a process; time-lapse video.

- Ask the family what's missing?

- What objects have been re-used for function or task different than its original purpose? Be sensitive to private activities (i.e. shower).

- Buy cameras ahead of time.

Once in the field, our team spent two weeks of intensive field research in Santiago, Chile, with families living in campamentos (slums). During the process, students developed and personalized the guided methodologies, making them their own.

Throughout the entire process, students documented their
research findings and created people's profiles and floor
plans with flows of water and daily activities.

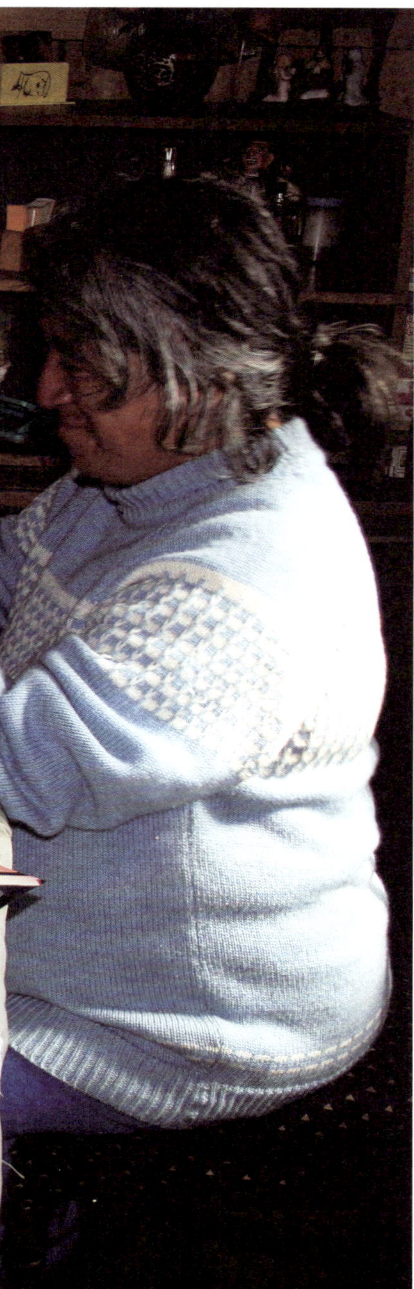

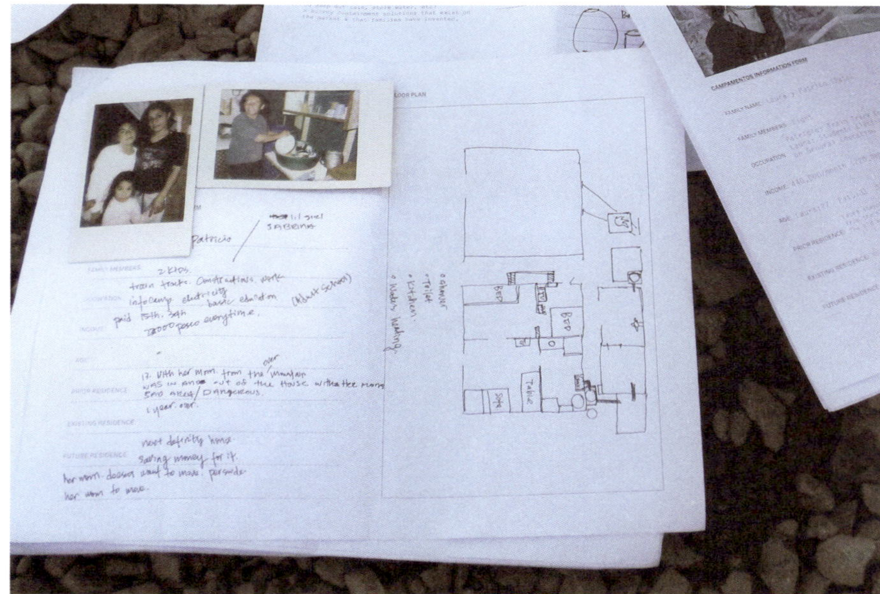

Students Diane and Ramon (above) look at the documentation
of the research findings they have just created while taking
a break in the field.

Student Jessica Yeh experiences firsthand what daily life is like without running water.

DESIGN PROCESS

By Dan Gottlieb and Penny Herscovitch

WHAT MADE THE SAFE AGUA design process unique? The class proceeded like any other Transdisciplinary Studio (TDS) at Art Center, but with four significant disctinctions: first, driven by field research, student designers became opportunity seekers; second, collaboration was essential to the process; third, the studio embraced the ingenuity and resourcefulness embodied by Un Techo's *mínimo* philosophy; and finally, the resulting projects are designed to be rapidly implemented in the real-world campamento context.

DRIVEN BY FIELD RESEARCH: FROM PROBLEM SOLVERS TO OPPORTUNITY SEEKERS

During our research trip, each methodology card asked "How can we best impact this area?" Upon returning to Art Center from Chile, this question became the driving force for the design process. After compiling the field research, we clustered the research insights into areas of focus, ranging from long-term well-being (health, employment, education, and emotion), to daily water-related tasks and functions (optimizing containing, transporting, and using water for time efficiency and physical convenience).

Unlike many studio classes, in which an instructor or partner company might assign a project brief that defines the problem to tackle for the term, each Safe Agua team embarked on a process to define the problem for themselves, based on the families' needs identified during field research. This shifts the conventional responsibility of the design student to engage in the process of evaluating *which* (of the dauntingly many) problems to tackle, and then deciding *what* to design.

"What is *the* problem?" probed visiting faculty Adlai Wertman and Abby Fifer Mandell from the Society and Business Lab at USC's Marshall School of Business. While the constellation of daily and long-term challenges that people in the campamentos face seemed daunting, intertwined, and complex to us, Adlai's provocation challenged

each team to focus on a very specific problem that could be tackled in the remaining 10 weeks.

At one point during this process of problem definition, Environmental Design student Stephanie Stalker asked, "Rather than identifying potential problems to solve, couldn't we identify opportunities?" Although it might seem a simple question of semantics, Stephanie's question shifted our view of the problem-solution paradigm toward a much more optimistic perspective: we may have begun by calling ourselves problem solvers, but in fact we would come to define ourselves as opportunity seekers.

COLLABORATION: DESIGNING "WITH" NOT "FOR"

Early in the term, Dirk-Mario Boltz, visiting professor from the Berlin School of Economics, framed Safe Agua in terms of the larger "co-" trend: co-creation, collaboration, etc. At heart, the Safe Agua design process embraced collaboration — an engagement between people of different disciplines, perspectives, and histories.

Our ongoing dialogue with and responsibility to the campamento families and our NGO partner, Un Techo, drove the project. Co-creation, simply put, is designing *with* people, not *for* them. The first step is empathy — moving past a mindset of "us" and "them" to a mindset of "we."

In Product Design student K.C. Cho's words, "No statistic or data replaces direct contact and feedback from the families. Once we were able to connect with the families, they gave us everything we needed to start the project."

In practice, bridging the divide of location, culture, and language presented challenges. Yet, students stayed directly connected to the families by several means, including email and Skype. Families from the campamento participated in a focus group organized by Un Techo, to share their specific feedback on each project. This dialogue between students and families extended to co-testing: as Jessie and Narbeh tested their *Ducha Halo* shower prototype in Pasadena, families were testing it in the campamento. Since the campamento families could not come to Pasadena for the final presentation, our team sent the families a five-foot long banner of the entire final presentation, with all of the students' work. This symbolic exchange extended the personal relationships established during the project.

MÍNIMO: MAXIMUM IMPACT WITH MINIMAL RESOURCES

Mínimo stands for maximum impact for minimal resources — this is the philosophy of Un Techo's Innovation Center. *Mínimo* also encompasses the extraordinary resourcefulness and ingenuity that people living in the campamentos have developed out of necessity. The Safe Agua studio adopted the *mínimo* ethos — that design innovation can be driven by a radically low budget. In practice, this approach influenced every aspect of the studio, from process to presentation to final prototypes. For midterm, teams employed an iterative process of making working mock-ups, "Frankensteined" together from off-the-shelf parts (rather than more polished looking but non-functional models). This iterative process of making full scale, working mock-ups continued through to the final, to yield final prototypes intended for real-world implementation.

REAL-WORLD DESIGN FOR THE CAMPAMENTO CONTEXT

The context of the campamento was paramount, and communicating the context in which each design was intended to become a part of the challenge. A particularly inventive student team played a Spanish soap opera on high volume during their research presentation to help the guests understand what it was like inside the homes of many of the campamento families. For midterm and final presentations, the simple rule of "no display pedestals" pushed teams to display their proposals within a context that communicated the feeling of the campamento to the whole school; and teams displayed their projects in the gallery amidst decidedly un-gallery-like wooden slat structures. You cannot remove these projects from their context; they simply do not make sense against a glowing white backdrop, as they are for, by, and of the campamento.

Part of our collective responsibility as a class was to bring the *mínimo* ethos, and our connections with the families of the campamentos, back to Art Center, to share with the school and beyond. This is precisely what this book endeavors to do: to connect you with the people in the campamentos, the challenges they face, the bigger picture of global water and poverty challenges, and the solutions proposed by Safe Agua teams.

A young girl helps her mother carry
empty bottles for washing dishes.

A Spontaneous Farm / Infocap is a technical school for adults that is located near the offices of *Un Techo para Chile*. At the school, there is a beautiful garden where chickens, goats, and peacocks happily coexist with plants, vegetables, and even sheep. The garden is full of life, and visitors to Un Techo always get the impression that someone has been taking good care of it for many years. / The garden originated as a gift from a florist to Father Felipe during times of drought. The flower-growing business was not going well; because of the drought, plants were dying. There was not enough water to keep them alive, so the florist had less and less work each day. Due to her difficult situation, the garden became a burden for her and she didn't know what to do with it anymore. Then she had the brilliant idea of giving the garden to Father Felipe, who didn't know what to do with it either! / As time went by, the friends who visited Father Felipe began to think that the garden was looking a bit sad. One fine day somebody brought him a sheep as a gift so that the garden wouldn't be so empty anymore. Soon after, another friend thought that the sheep looked lonely in the otherwise empty garden, and decided to give Felipe a small goat to keep it company. Then, over time, people coming to Infocap assumed that Felipe was starting a farm, and many decided to help him out by bringing chickens and peacocks, and by planting different kinds of vegetables. Before long, and without Father Felipe trying to make it so, there were many animals living a happy life together, and the flower garden had, in fact, turned into a farm. / When we visited Father Felipe, and remarked on the beautiful little farm, he mentioned that there was no longer room for so many animals. There were already three sheep, and not enough grass for grazing. When we asked Felipe what was he planning on doing to improve the situation, he smiled, tapped his hand on his belly and said, "We will have a good meal!"

By **Felipe Berríos**, Founder of *Un Techo para Chile*,
as told to **Liliana Becera**, Faculty, Product Design,
Art Center College of Design

THE SIX TRANSDISCIPLINARY TEAMS designed innovative solutions at a range of scales — from product to system, to community spaces to campaign — to address specific water-related needs identified through their field research. The teams worked together to coordinate how each proposal could complement one another, and produce a result where the whole is greater than the sum of its parts.

To kick off the project, Julián Ugarte, director of Un Techo's Innovation Center, envisioned the metaphor of the class as a human body. *Gota a Gota* is the heart of Safe Agua — a gravity-fed system that allows water to flow to all parts of the home. *Agua Segura* fulfills the physical need to drink, and *Ducha Halo* and *ReLava* fulfill the need to be clean. In addition, people need to communicate and to share social support — needs addressed by the *Mila* community laundry and the *Íindex* catalog of shared innovations.

The projects that emerged are interconnected by shared aspirations. While each project specifically targeted water-related challenges, the entire class worked toward a larger, holistic goal, to make an impact on the lives of families in the campamentos.

The projects are also interconnected in a pragmatic sense, in that they collectively fit into the home, to make incremental improvements in the quality of daily life. More broadly, they fit into the longer trajectory of someone's life, and seek to help make the transition from people's current temporary living situation in the campamento as they move toward a better life for their families in the *casa definitiva* (permanent social housing).

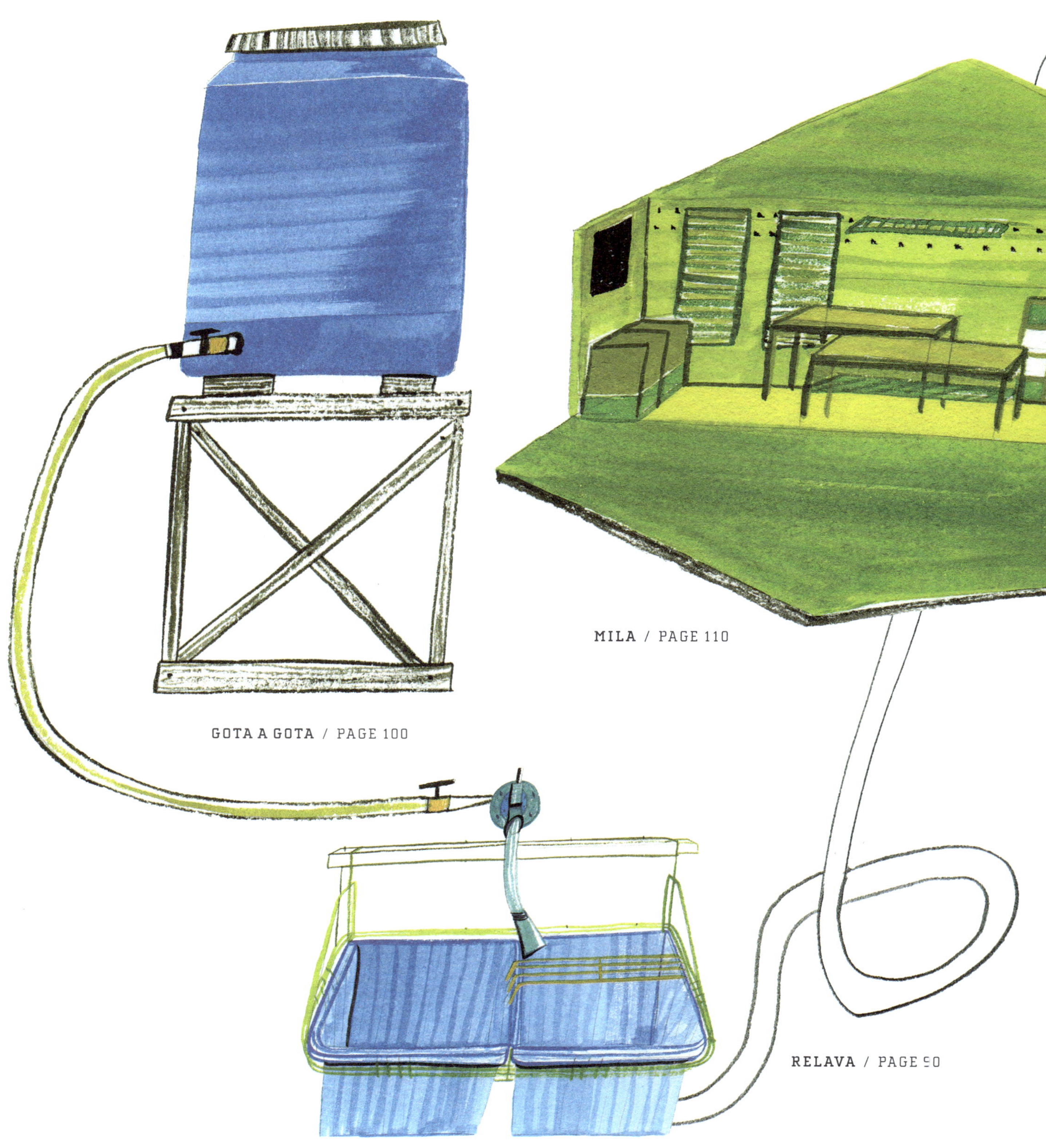

GOTA A GOTA / PAGE 100
MILA / PAGE 110
RELAVA / PAGE 90

iindex

Karen and Sonia playing in Campamento San José.

"TODAY 2.5 BILLION PEOPLE, INCLUDING ALMOST 1 BILLION CHILDREN, LIVE WITHOUT EVEN BASIC SANITATION."

SOURCE | Water Supply and Sanitation Collaborative Council (WSSCC).

A young boy tests out a *Ducha Halo* prototype and enjoys a warm shower.

DUCHA HALO

DESIGNED BY NARBEH DEREGHISHIAN AND JESSICA YEH

Ducha Halo is a portable shower that ingeniously repurposes existing inexpensive parts to bring the dignity and well-being of a hot shower to people living with no running water.

BATHING "BY PARTS" (pouring a can of water on each body part) is not a dignified or healthy way to shower. *Ducha Halo* provides people in the campamento with a real shower. This inexpensive portable shower can be easily assembled from $17 worth of existing materials currently available at a hardware store. Two minutes of hand pumping yields a warm shower that lasts 15 minutes.

HOW DUCHA HALO IS BUILT

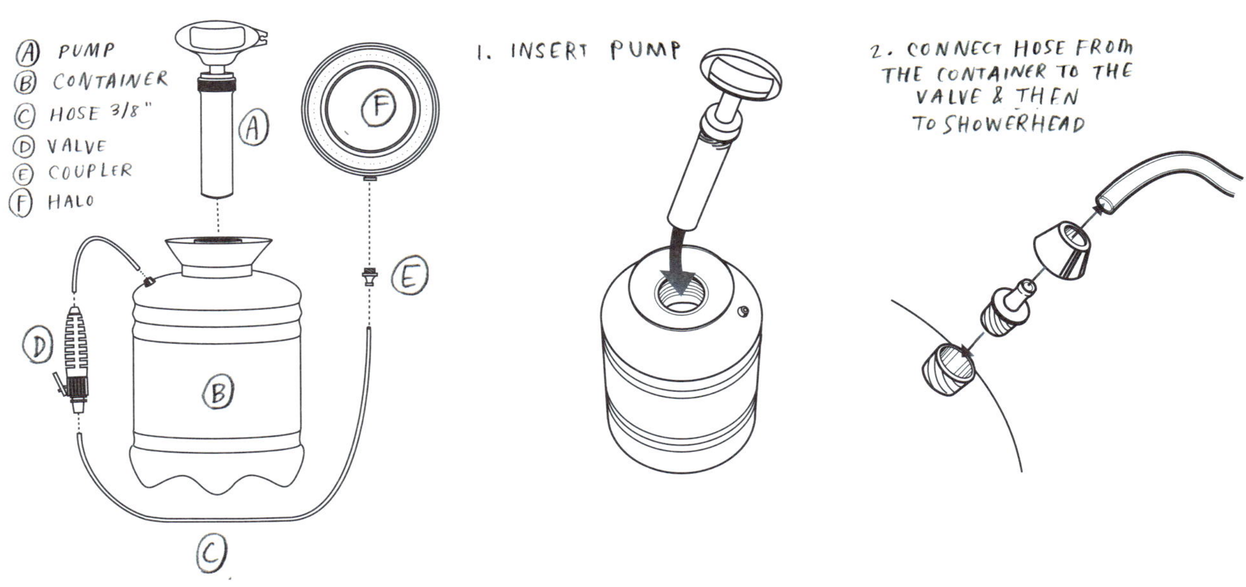

3. TIE DOWN VALVE SO THAT IT MAY BE CONTROLLED USING YOUR FEET

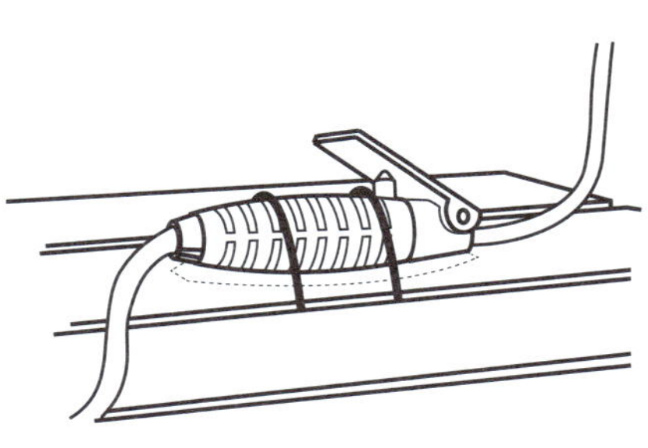

4. JOIN HALO SHOWERHEAD TO COUPLER & HOSE

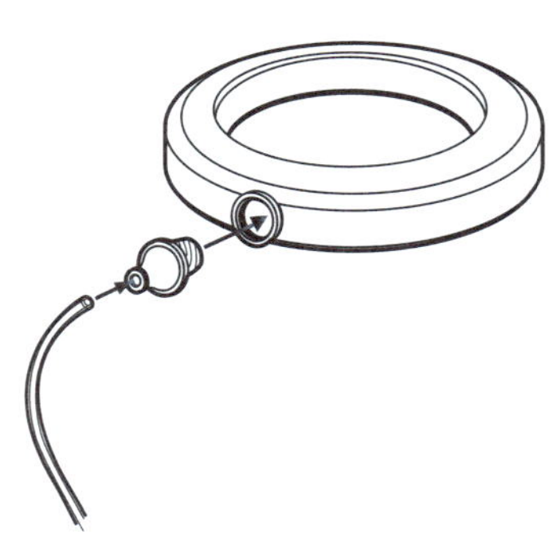

5. HANG HALO SHOWERHEAD

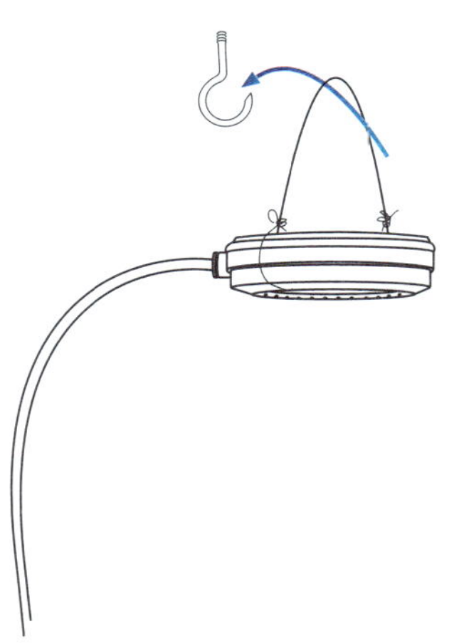

DUCHA HALO IS DESIGNED TO BE IMPLEMENTED IN TWO PHASES—A DIY KIT, FOLLOWED BY A PRODUCTION MODEL.

PHASE 1: DIY Kit

Un Techo is currently implementing the first phase in the campamentos, a DIY kit easily assembled entirely from $17 of existing parts from any hardware store.

PHASE 2: Production Model

The second phase comprises a production model, reengineered and designed to address the specific needs of people in the campamento in response to prototype testing and feedback. The six-liter metal container (with an analog temperature sensor) is easy to heat on the stove, reducing preparation time and physical strain.

80% OF FAMILIES IN THE CAMPAMENTO WASH BY PARTS.

IN CHILE'S CAMPAMENTOS, bathing is an involved process, requiring multiple steps, including heating, mixing, carrying and scooping water. Most families resort to bathing by parts: they fill up a bucket with a mixture of boiling and cold water, and use a small container to rinse themselves part-by-part. This process is time-consuming, cumbersome, requires several steps, and can cause physical strain and illness.

SOURCE | *Un Techo para Chile*, based on 20 families surveyed in Campamento San Pablo.

"I come back thirsty from work, eager to take a bath, but I have to pick up a can with some water, pick up the soap and the shampoo and try to wash myself inside the latrine. There is no dignity for anyone in this. I look forward to taking a proper shower. I dream of coming home from work and feeling the soap and flowing water. I think we should at least have the chance to take a decent shower."

PEDRO PASTEN, CAMPAMENTO DWELLER, SANTIAGO, CHILE

This is what a warm shower looks like at Cecilia's house.

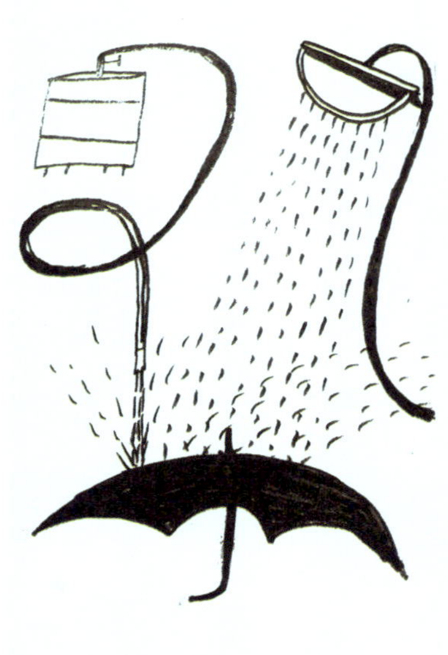

1 SHOWER PROTOTYPES WERE MADE FROM PUMPS, BEER KEGS TO LAWN SPRAYERS.

2 JULIAN TESTS THE DUCHA HALO DURING MIDTERMS AND CONCLUDES THAT IT FEELS LIKE A REAL SHOWER.

3 ROSITA CHOSE THE HALO SOLUTION BECAUSE IT WAS HANDS FREE.

4 TECHO RETURNED TO CHILE WITH A WORKING PROTOTYPE FOR FAMILIES TO USE AND GATHER FEEDBACK.

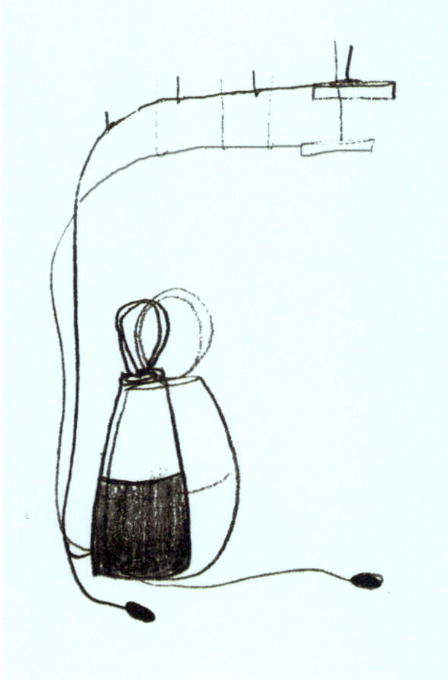

5 WITH FEEDBACK, JESSICA AND NARBEH REFINED DUCHA HALO TO SERVE THE EXACT PURPOSES OF CAMPAMENTO RESIDENTS.

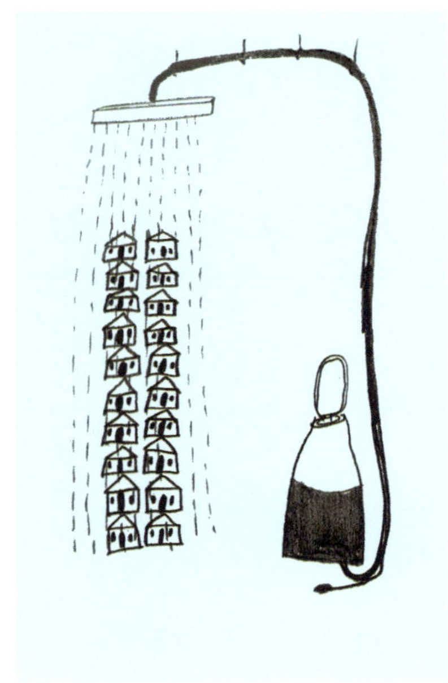

6 MEANWHILE, 20 FAMILIES SPENT 3 DAYS EACH TESTING THE DUCHA HALO.

7 CONCURRENTLY, JESSICA AND NARBEH BOTH SET UP THE DUCHA HALO IN THEIR OWN SHOWERS FOR 3 WEEKS.

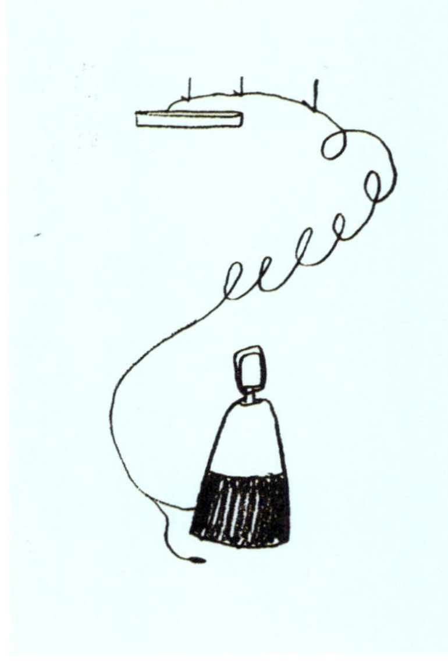

8 "ONE MILLION SHOWERS BY NEXT YEAR. THIS IS OUR PROMISE."

TESTIMONIALS

AT THE FINAL SAFE AGUA presentation, Julián exclaimed "One million showers by next year!" and indeed, *Un Techo para Chile* moved very quickly toward prototyping and implementing the *Ducha Halo*. Just as the 20 families in Campamento San José were completing testing the intial *Ducha Halo* shower prototype, an 8.8-magnitude earthquake struck Chile, on February 27, 2010. After the earthquake, the need for emergency relief products like the *Ducha Halo* portable shower became all the more acute, as the quake severely damaged more than 500,000 homes, affecting 1.5 million people.

In response to this natural disaster, Un Techo embarked on an ambitious and rapid response, to construct 30,000 *media aguas* as emergency housing. After the quake, Un Techo also moved quickly to assemble a refined prototype of *Ducha Halo* from parts locally available at the Chilean hardware store chain Sodimac. The Un Techo team brought this prototype to families living in emergency housing in the city of Constitución after losing their homes in the earthquake. The *Ducha Halo* prototype brought a real shower to these families and the results of this new version of the *Ducha Halo* in the earthquake zone were "un Boom" — a big hit!

The process of creating, testing, and refining the *Ducha Halo* in collaboration with the residents of Campamento San José and those affected by the earthquake embodies what renowned social innovation thinker C.K. Prahalad describes as "a new respect for consumers as co-creators of solutions and not just passive recipients of a product or service."

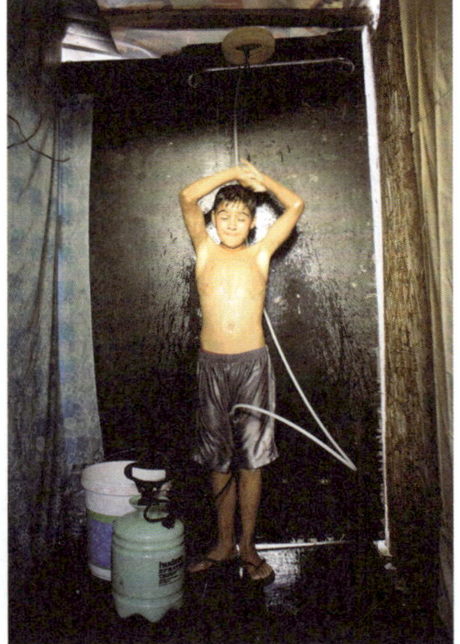

A young boy testing the *Ducha Halo* in campamento. Pumping for two minutes allows 15 minutes of waterflow.

Transfering water from the storage
barrel to another container, for
washing dishes outdoors.

"I DON'T LIKE TO DO THE DISHES IN FRONT OF THE HOUSE BECAUSE I DON'T WANT PEOPLE TO SEE ME." MARIA, CAMPAMENTO SAN JOSÉ, SANTIAGO, CHILE

A woman in the campamento tests the functional *ReLava* prototype manufactured by a Chilean wire company.

RELAVA

DESIGNED BY JACQUELINE BLACK AND K.C. CHO

An inexpensive kitchen workstation that makes it sanitary, efficient, and dignified to wash dishes indoors and to reuse of water.

RELAVA INTEGRATES A SINK for soaking, rinsing, and cleaning dishes, a drain for reusing water, and a drying rack into a single kitchen workstation. *ReLava's* design repurposes plastic tubs into sink basins, held by a collapsible wire frame that hangs from the wooden wall of Un Techo's transitional houses. With the *ReLava* frame, it is simple to convert standard plastic tubs into a working sink with a drain by drilling a hole in the tub with a hand-held "hole-saw" and affixing a hose.

RELAVA'S DRAIN FACILITATES the reuse of water, as women in the campamentos are accustomed to reusing dishwater to flush the latrine or to keep dust and insects away. Embedded in the *ReLava* proposal is the opportunity for people to generate income by converting existing bins into sinks, and selling *ReLava* frames and bins at the *feria* (the local market where many campamento residents both shop and sell their wares). *ReLava* exemplifies Un Techo's *mínimo* ethos, creating maximal impact with minimum resources. Washing dishes in the campamentos is an arduous task. Lacking access to running water and drainage, the simple act of washing dishes requires numerous steps, carrying and moving water by hand to complete the task.

Without a fixed workstation, women often wash dishes on the floor, in unsanitary conditions, or outdoors subject to the weather.

Because there is no running water, families in the campamento use water very efficiently and are conscious of exactly how much water they use. When families move to permanent social housing, with the support of Un Techo, they will have running water, indoor plumbing, and convenient kitchen workstations, but they will be also be confronted with the reality of paying for water for the first time. *ReLava* is a first step in shifting from ad hoc use of water in the *media agua* to help transition to the formalized use of water in their future permanent homes.

1. Maria carries water in small containers from the large storage barrel to do dishes. 2. Mireya sets up a tub to do dishes on the floor with her daughter. 3. Maria uses two plastic buckets on top of a surface outdoors to soap and rinse dishes. 4. Maria does dishes balanced on two wooden boards, atop an old toilet basin to catch the water. 5. Mireya carries used dish water outside and pours it on the ground. 6. Noelia carries buckets of grey water to use for the latrine.

THE RELAVA DEVELOPMENT PROCESS

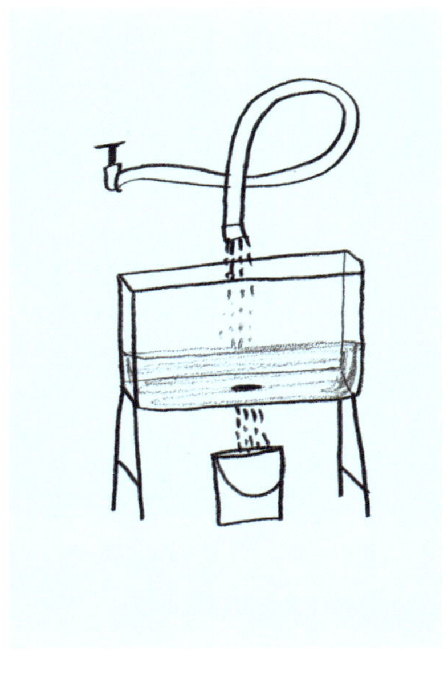

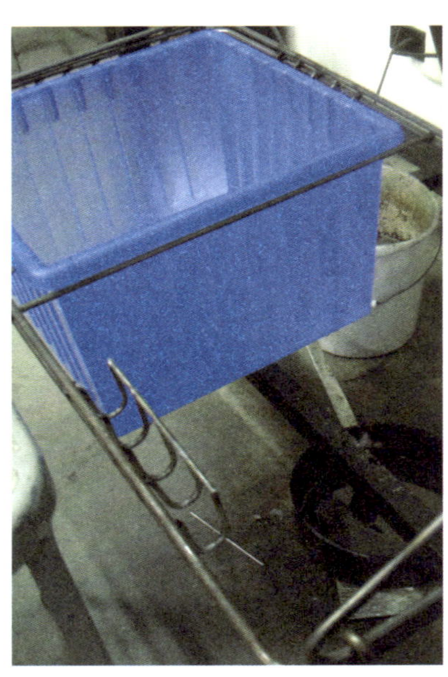

1 THE LACK OF RUNNING WATER PROLONGS AND AGGRAVATES EVEN THE SMALLEST HOUSEHOLD TASKS.

2 FAMILIES SAID THEY WANT ONE PRODUCT THAT HAS WASH + RINSE + CLEAN, AND A DRAIN.

3 A WIRE MANUFACTURER IN CHILE WHO WORKED WITH TECHO HELPED KC & JACKIE DESIGN A WIRE BASED SINK.

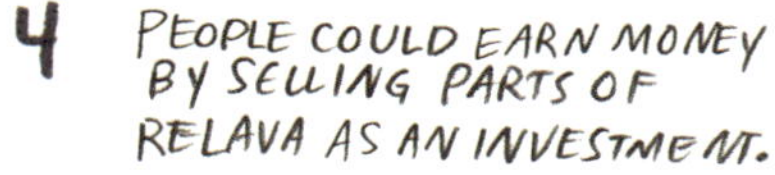

4 PEOPLE COULD EARN MONEY BY SELLING PARTS OF RELAVA AS AN INVESTMENT.

5 JACKIE AND K.C. PROTOTYPE A WIRE-FRAME SINK TO BE MADE BY A CHILEAN WIRE MANUFACTURER.

6 TWO TUBS WITH A SELF DRILLED HOLE AND A WIRE FRAME CAN BE CARRIED ONTO A BUS IN UNDER A MINUTE.

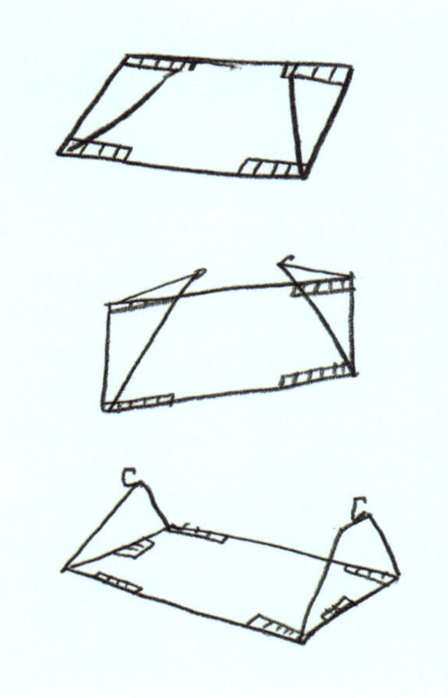

7 MARIA IS ABLE TO UNFOLD AND HOOK HER RELAVA ONTO THE MEDIA AGUA WALL.

8 USED WATER DISPENSES UNDERNEATH THE MEDIA AGUA TO KILL INSECTS AND REUSED FOR THE TOILET.

CONCLUSION

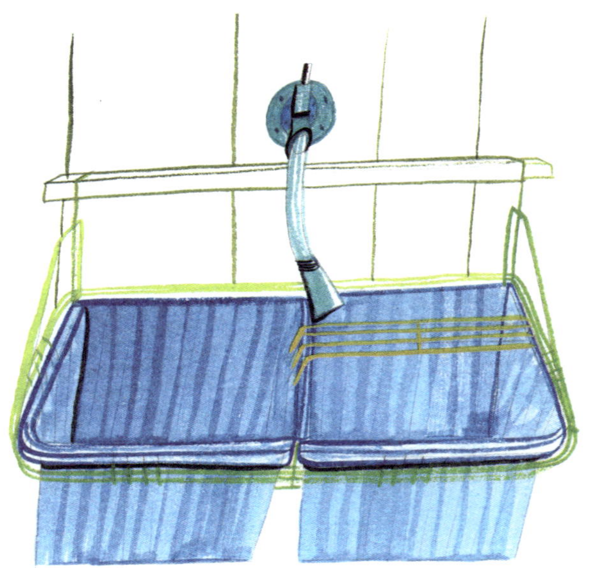

AFTER THE 8.8 MAGNITUDE February 2010 earthquake struck Chile, 1.5 million people's homes were severely damaged. These families found themselves living in campamentos and emergency relief housing for the first time, without the daily comforts and facilities that they had been accustomed to their whole lives — including kitchen sinks and running water. For families living in earthquake relief *media aguas*, the *ReLava* sink is an ideal way to quickly set up an inexpensive, sanitary emergency kitchen sink and workstation, which can unfold and hang from the *media agua's* wooden wall. Immediately after the earthquake hit, Un Techo worked with the Chilean wire manufacturer, Inchalam-Acmanet, to manufacture and test a refined prototype of *ReLava's* wire frame. During the testing, common plastic tubs fit perfectly into the wire frame. Testing in the factory and in Campamento San José confirmed that the *ReLava* metal frame, hung from the wooden wall, was able to securely hold 30 kilograms of water — with the sink's two 16-liter basins filled with water.

LEFT: Testing the refined, Chilean-manufactured *ReLava* workstation prototype in Campamento San José. ABOVE: *ReLava* wire frames and plastic basins can be distributed by vendors at the local *ferias* or informal markets.

Carrying water from bucket to bucket throughout the home requires tremendous effort, energy, and time.

" I HAVE TO CUT OPEN LARGE CONTAINERS
IN ORDER TO SCOOP WATER AND TRANSPORT IT.
IT'S THE ONLY OPTION AND I HATE IT.
IT'S IMPOSSIBLE TO KEEP
WATER CLEAN. " ROSA, CAMPAMENTO
SAN JOSÉ, SANTIAGO, CHILE

The lack of a drainage system
forces Maria, who has osteoporosis,
to physically discard the water.

GOTA A GOTA

DESIGNED BY STELLA HERNANDEZ,
NUBIA MERCADO-RAMIREZ AND DIANE JIE WEI

"If I had the money, I'd buy a tap to simulate running water." MARIA, CAMPAMENTO SAN JOSÉ, SANTIAGO, CHILE

UN TECHO CHALLENGED THE TEAM to provide a solution to bring people in the campamentos running water, at the turn of a handle or the push of a lever. Through research into pressurized water systems, we found that harnessing gravity is the most efficient and affordable way to bring running water to different parts of the home, to alleviate the burden of moving large quantities of water by hand. While a few families in Campamento San José have built gravity fed systems, these systems still require women to lift water to fill the elevated storage tank by hand.

The *Gota a Gota* ("Drop by Drop") system harnesses gravity to afford families the ease, convenience, and dignity of turning on a tap to bring running water into the home.

GOTA A GOTA HELPS MAKE THE DREAM OF RUNNING WATER A REALITY.

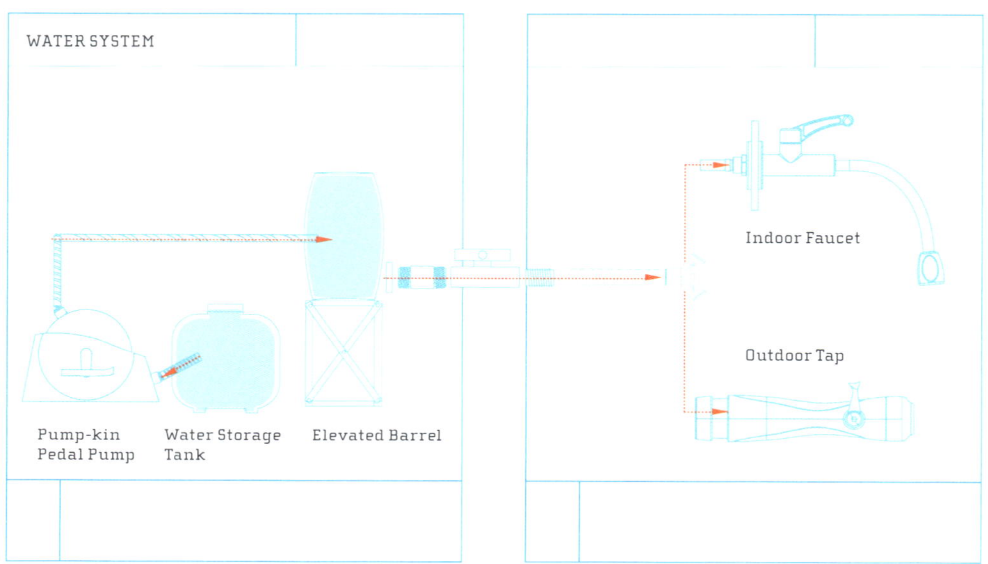

MANY OF US TAKE FOR GRANTED that water flows from the tap at the turn of a handle, but for people living in the campamentos, running water remains a dream. Without access to running water in their homes, people must carry large quantities of water from outdoor storage containers for daily tasks, such as laundry, washing dishes, cooking, and cleaning. Carrying water from bucket to bucket to bucket throughout the home requires tremendous effort, energy and time. This burden impacts health and well-being, especially for women, who are often the ones responsible for managing water to care for their families.

An indoor faucet with a flexible nozzle
and an easy on-off handle shortens the
length of dishwashing time and allows
for better control of water distribution.

<u>GOTA A GOTA</u>

THE STORAGE TANK SITS ATOP
A SELF-CONSTRUCTED WOODEN
PLATFORM (WHICH THE GOTA
A GOTA MANUAL SHOWS HOW
TO CONSTRUCT.)

<u>PUMP.KIN</u>

A FOOT PUMP DESIGNED
TO ELEVATE WATER TO AN
EXTERIOR STORAGE TANK.
PUMP.KIN IS EASY FOR WOMEN,
KIDS & ELDERLY TO PEDAL,
STANDING UP OR SITTING DOWN.

OUTDOOR TAP

WATER FLOWS FROM THE ELEVATED STORAGE TANK, THROUGH A HOSE, TO AN ERGONOMIC FAUCET WITH A CONVENIENT CLIP FOR TASKS PERFORMED OUTDOORS, SUCH AS LAUNDRY.

INDOOR FAUCET

ANOTHER HOSE LEADS INTO AN INDOOR FAUCET, WITH A FLEXIBLE NOZZLE AND EASY ON-OFF HANDLE, TO MAKE WASHING INDOORS COMFORTABLE AND CONVENIENT.

TESTIMONIALS

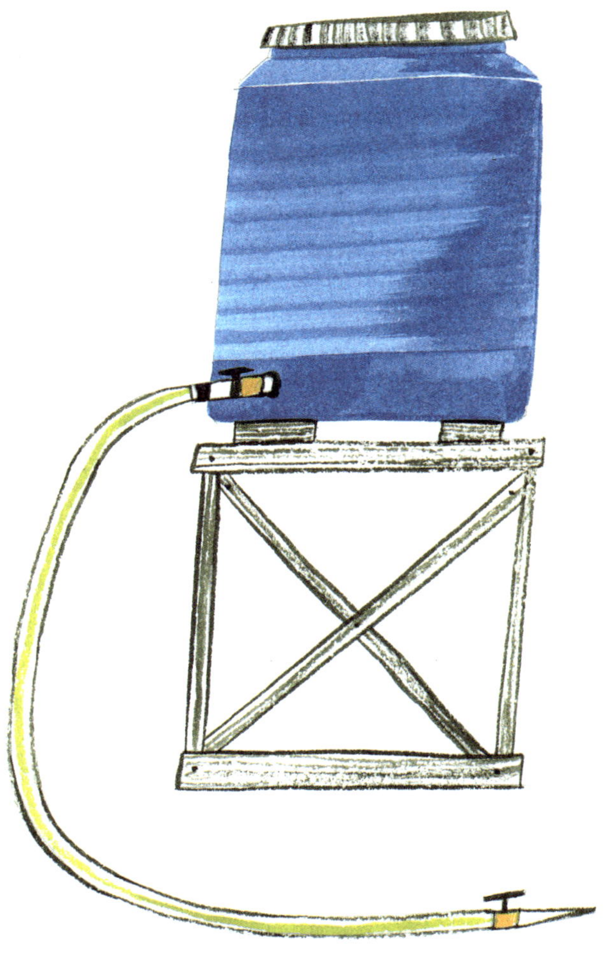

"This project makes me understand how design can change people's life, but most importantly, my social responsibility as a designer."

DIANE JIE WEI, PRODUCT DESIGN STUDENT

"Safe Agua gives us an opportunity to implement our design training to the other 90% of the world that is usually disregarded. As design students, we often get carried away with wanting to design glamorous objects; in this case, we are designing solutions that will really make an impact on somebody's life. This project has opened my eyes to see a much broader need for design in the world."

NUBIA MERCADO RAMIREZ, TRANSPORTATION DESIGN STUDENT

A resident pours water into a cut-open storage barrel outside her home. She simulated her own running water system by attaching a line from the barrel into her home.

Maria prepares to do laundry for her family.
On a hot day, the entire process takes about six
hours of labor, but during the winter, clothes
can remain wet for up to two weeks.

WOMEN AND <u>CHILDREN</u> BEAR THE BURDENS DISPROPORTIONATELY, OFTEN SPENDING <u>SIX HOURS</u> OR MORE EACH DAY FETCHING WATER FOR THEIR FAMILIES AND COMMUNITIES.

SOURCE | http://www.blueplanetrun.org

Because of the shortage of water, laundry water is reused multiple times.

MILA

Mila is more than a space to do laundry, it's a gathering space for the entire community.

LAUNDRY IS THE HOUSEHOLD TASK that requires the most effort, time, and energy. The 20 families living in Campamento San José collectively spend over 500 hours a week on laundry. What if, for the price of one bottle of Fanta, each family could do a month's worth of laundry in half the time, without carrying or lifting any water? _Mila_ stands for _mi lavandería_: a vibrant community laundry facility that offers access to washing machines to save women time and reduce injury in the washing process, and creates opportunities for women to earn income.

LAUNDRY IS THE HOUSEHOLD TASK IN THE CAMPAMENTO THAT REQUIRES THE MOST TIME AND EFFORT, AND CAN CAUSE INJURY AND ILLNESS.

IT BECAME APPARENT early in our research that laundry is the one household task we really needed to address. Without running water, doing laundry by hand or repeatedly scooping water into often-broken machines not only takes time, but also causes task-related injuries such as swollen joints, carpal tunnel syndrome, and back problems. During winter, washing clothes outside in the cold and hanging clothes to dry in the home can lead to pneumonia in both women and kids.

What is it like for women in the campamentos to do laundry by hand or with broken machines now? The whole process takes at least six hours, and as many as 12. And laundry day is repeated two to three times a week.

1. Laundry starts with carrying a lot of water in buckets from the storage barrel to a small washing machine or basin, or even by carrying the washing machine outside. 2. Scrubbing, rinsing, and wringing out water from clothes by hand with cold water can lead to task-related injuries such as swollen joints and carpal tunnel syndrome. 3. Bending to fill and empty machine or tub can lead to back pain. Clothes take a long time to dry out in the cold, and hanging clothes to dry in the home in winter can make the kids sick.

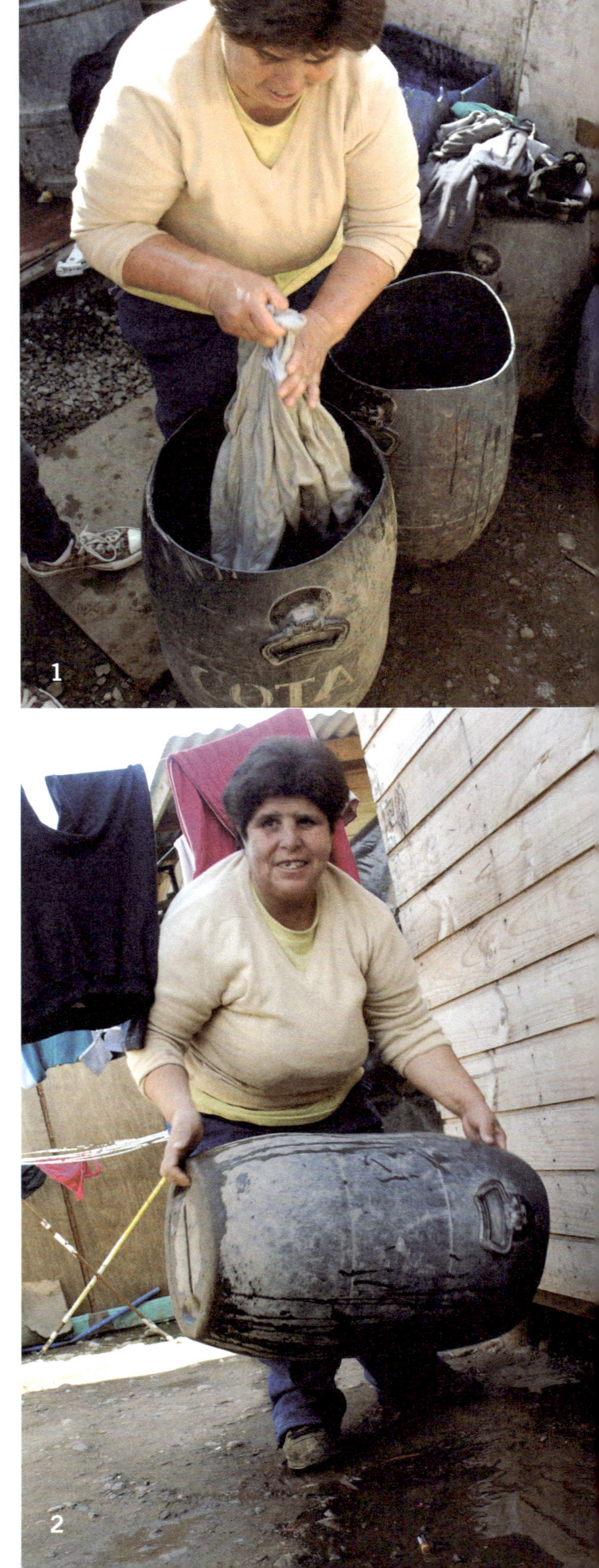

Mila

FROM HANDWASHING CLOTHES
TO A COMMUNITY LAUNDRY

PLANNING & BUILDING
MILA COMMUNITY LAUNDRY

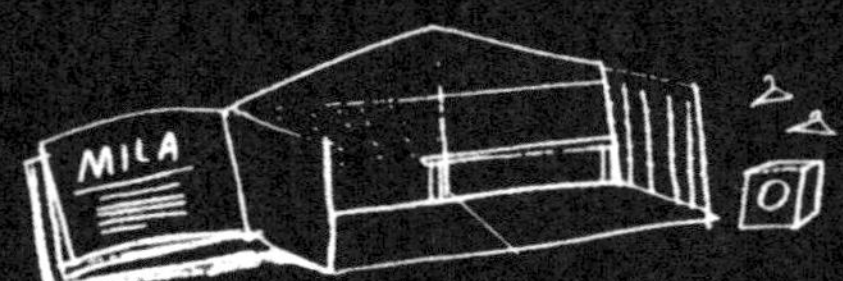

① CAMPAMENTO RESIDENTS WILL USE THE MILA HANDBOOK TO PLAN WHAT THEIR FACILITY NEEDS ARE.

② ONCE THE COMMUNITY PLANS MILA FOR THEIR CAMPAMENTO, TECHO WILL TEAM UP WITH A SPONSOR COMPANY SO THE COMMUNITY BENEFITS FINANCIALLY.

③ MILA IS BUILT IN A NEW MEDIA AGUA!

USING MILA

ONCE MILA IS BUILT, RESIDENTS WILL WALK A FEW METERS TO MILA AT A SCHEDULED TIME TO DO LAUNDRY.

A COMMUNITY SUPERVISOR WHO LIVES IN THE CAMPAMENTO WILL HELP RUN MILA.

A FIX-IT PERSON WILL RECEIVE JOB TRAINING TO KEEP MILA RUNNING SMOOTHLY.

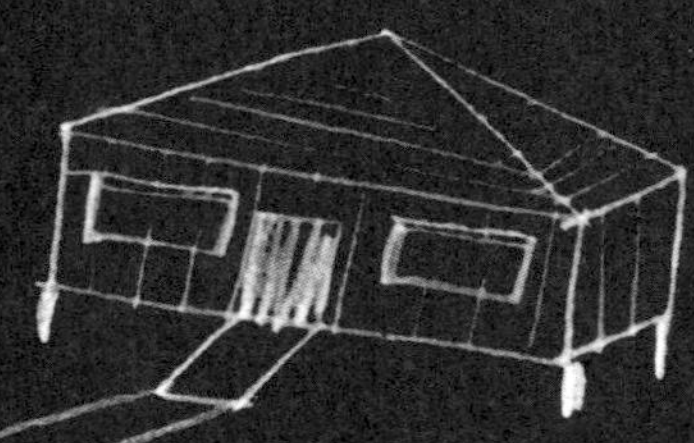

RESIDENTS CAN ALSO EARN INCOME BY DOING LAUNDRY FOR PEOPLE OUTSIDE THE CAMPAMENTO.

AT THE CASA DEFINITIVA, RESIDENTS CAN APPLY FOR A MICRO-LOAN & OPERATE THEIR OWN LAVENDARIA ON THE GROUND FLOOR.

CONCLUSION

"My passion for this project lies in the idea of community — fostering interaction and building relationships. Designers are problem solvers. Involving creative minds and problem solvers in any social entrepreneurial project opens up the opportunity to address and/or solve issues and improve quality of life, but also to inspire and give hope."

STEPHANIE STALKER, ENVIRONMENTAL DESIGN STUDENT

UN TECHO is working to partner with companies such as Unilever to make *Mila* community laundry facilities a reality for people.

Stephanie's final model of *Mila*,
a proposal for a community laundry
for the campamentos.

A young girl dashes past a row of empty
plastic bottles waiting to be filled with
water from the water delivery truck.

ONE-IN-SIX PEOPLE IN THE WORLD

LACK SAFE DRINKING WATER.

EVERY 15 SECONDS, A CHILD

DIES FROM A WATER-RELATED DISEASE.

SOURCE | http://www.blueplanetrun.org

Maria drinks water after it has been
boiled, but admits to forgetting to
treat her water from time to time
because it is a hassle.

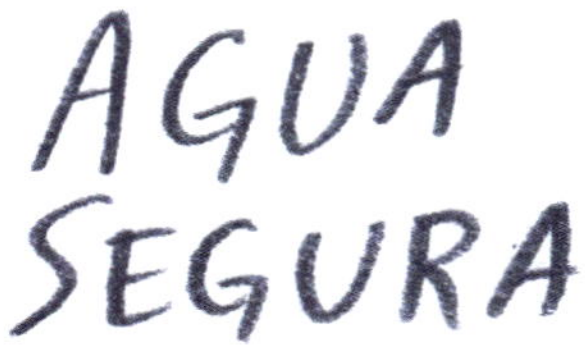

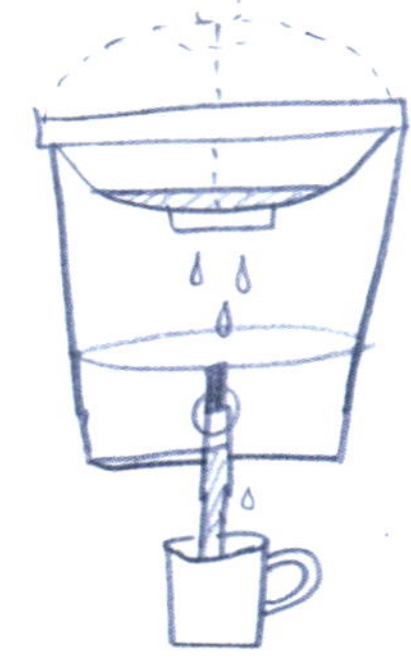

DESIGNED BY ELIZABETH BAYNE AND ERICA LI

A family-sized kit for water chlorination and filtration to ensure safe, easy, pure water for drinking and cooking.

OUR TEAM'S CHALLENGE was to address the issue of safe drinking and cooking water to prevent disease. Water treatment has been tackled in many different ways, by many different groups, but the challenge was in solving it in the context of the campamentos in Santiago, Chile. We found that existing solutions for subsistence conditions weren't quite appropriate, as residents had access to municipal water. But high-end solutions used in developed urban areas also didn't fit due to a lack of infrastructure and running water. We had to design a water treatment product for people in the middle: people who have infrequent access to water and store it for several days at a time.

Maria treats her dishwashing water with chlorine.

A SINGLE DROP OF WATER is exposed to numerous contaminants, from the point of collection (leaving the fire hydrant), to delivery by truck, to storage in a barrel (which can be contaminated by animals, kids, germs, and dirt), to scooping and carrying in a smaller container, before the moment when someone drinks it.

How do campamento residents currently treat water that may have been contaminated en route to the *media agua* home or during storage? Either by boiling or adding chlorine. However, these treatment methods are used infrequently, if at all. Boiling takes time and energy. Measuring chlorine accurately is difficult and inconvenient.

AGUA SEGURA IS AN EASY TO USE PURIFICATION & FILTRATION KIT THAT FITS ONTO ANY STANDING 5-GALLON BUCKET

Chlorine is a cheap, reliable way to treat water to prevent water-borne diseases like cholera, typhoid, and hepatitis A. In fact, nearly every utility in the United States uses chlorine to disinfect water. Chlorine is already part of the families' daily cleaning routine, and *Agua Segura's* built-in measuring device makes it easy and fast to accurately chlorinate water at the point-of-use.

A filter pack of activated carbon (produced through local sustainable business models) removes potentially harmful chlorine byproducts, and improves the odor and taste. A tap affixed to the bucket makes it easy to dispense the purified and filtered water into a glass. *Agua Segura's* design inspires confidence and promises safe and accurate water treatment.

THE *AGUA SEGURA* PROCESS

1 A SINGLE DROP OF WATER IS EXPOSED TO NUMEROUS CONTAMINANTS.

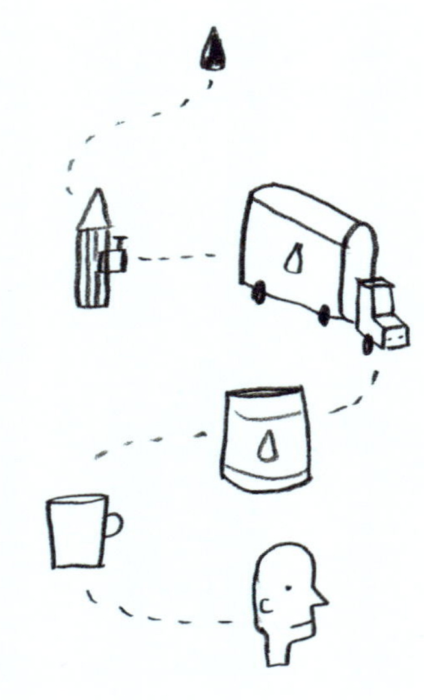

2 FROM LEAVING THE FIRE HYDRANT TO DELIVERY BY TRUCK, TO STORAGE IN A BARREL, TO THE MOMENT WHEN SOMEONE DRINKS IT.

3 THREATS TO WATER QUALITY INCLUDE: ANIMALS, CHILDREN, DIRT, GERMS.

4 CAMPAMENTO RESIDENTS CURRENTLY TREAT WATER INFREQUENTLY IF AT ALL.

5 CHLORINE IS A CHEAP, RELIABLE WAY TO TREAT WATER TO PREVENT WATER-BORNE DISEASES.

6 THE CLASS HAD PREVIOUSLY CONSIDERED THE USE OF CHLORINE BLEACH AS A NEGATIVE. BUT CAME TO SEE ITS POTENTIAL AS A LOW-COST WAY TO PURIFY WATER.

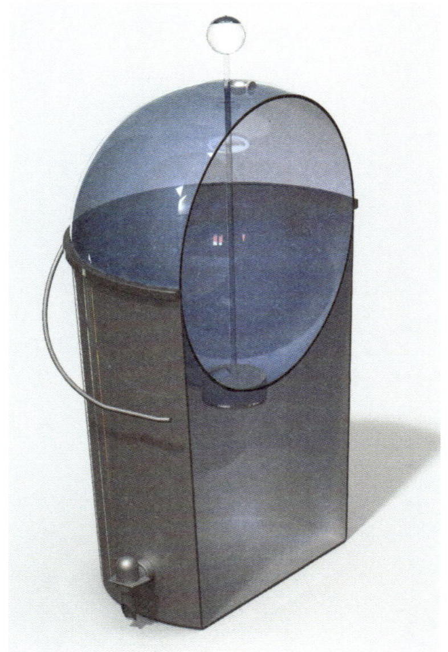

7 THE FAMILIES SAID THAT THEY NEEDED A WAY TO DISINFECT DRINKING WATER IN LARGE QUANTITIES AND ALSO MAKE IT TASTE BETTER.

8 AGUA SEGURA MAKES IT EASY TO MEASURE CHLORINE QUICKLY AND ACCURATELY TO PURIFY WATER AND FILTERS THE CHLORINATED WATER SO IT'S SAFE AND TASTES BETTER.

LARGE ENOUGH FOR AN ENTIRE FAMILY. AGUA SEGURA IS AFFORDABLE AND EASY TO MAINTAIN.

TESTIMONIALS

"This project is about making a contribution to what will soon be a global water crisis. It's a large issue in terms of scarcity, health, and sanitation. I have researched these issues in the past and this is an opportunity to develop actual solutions. I have a profound interest in water and identifying simple practices that drastically improve quality of life and reduce health risks."

ELIZABETH BAYNE, GRADUATE BROADCAST CINEMA STUDENT

"Other kids have gotten sick because of the water, but not mine — because we make sure we sanitize everything with chlorine."

MARIA PENALOZA, CAMPAMENTO SAN JOSÉ, SANTIAGO, CHILE

Children excitedly await
the water truck's arrival.

This discarded Coke bottle is now a boy's personal water dispenser in his tree house.

"SOME OF THE MOST CREATIVE INVENTIONS ARE BORN OUT OF THE COMBINATION OF NECESSITY AND INGENIOUS RESOURCEFULNESS."

LILIANA BECERA, FACULTY, PRODUCT DESIGN, ART CENTER COLLEGE OF DESIGN

Íindex de Innovación is a monthly newsletter that shares DIY solutions with communities within other parts of the country.

ÍINDEX DE INNOVACIÓN

DESIGNED BY RAMON CORONADO AND WILLIAM TANG

Inventions by people living in the campamentos.

PEOPLE LIVING IN CHILE'S CAMPAMENTOS are already ingenious problem-solvers out of necessity. *Íindex* proposes a communication strategy/platform to be implemented by Un Techo's Social Innovation Center. *Íindex de Innovación* empowers people in the campamentos to become innovators and to improve quality of life for people in their own communities and beyond. The *Íindex* strategy is comprised of a competition to inspire innovative solutions to challenges faced in the campamentos, and a monthly publication and open source Web site to enable people in the campamentos to share and learn from each other's innovative solutions.

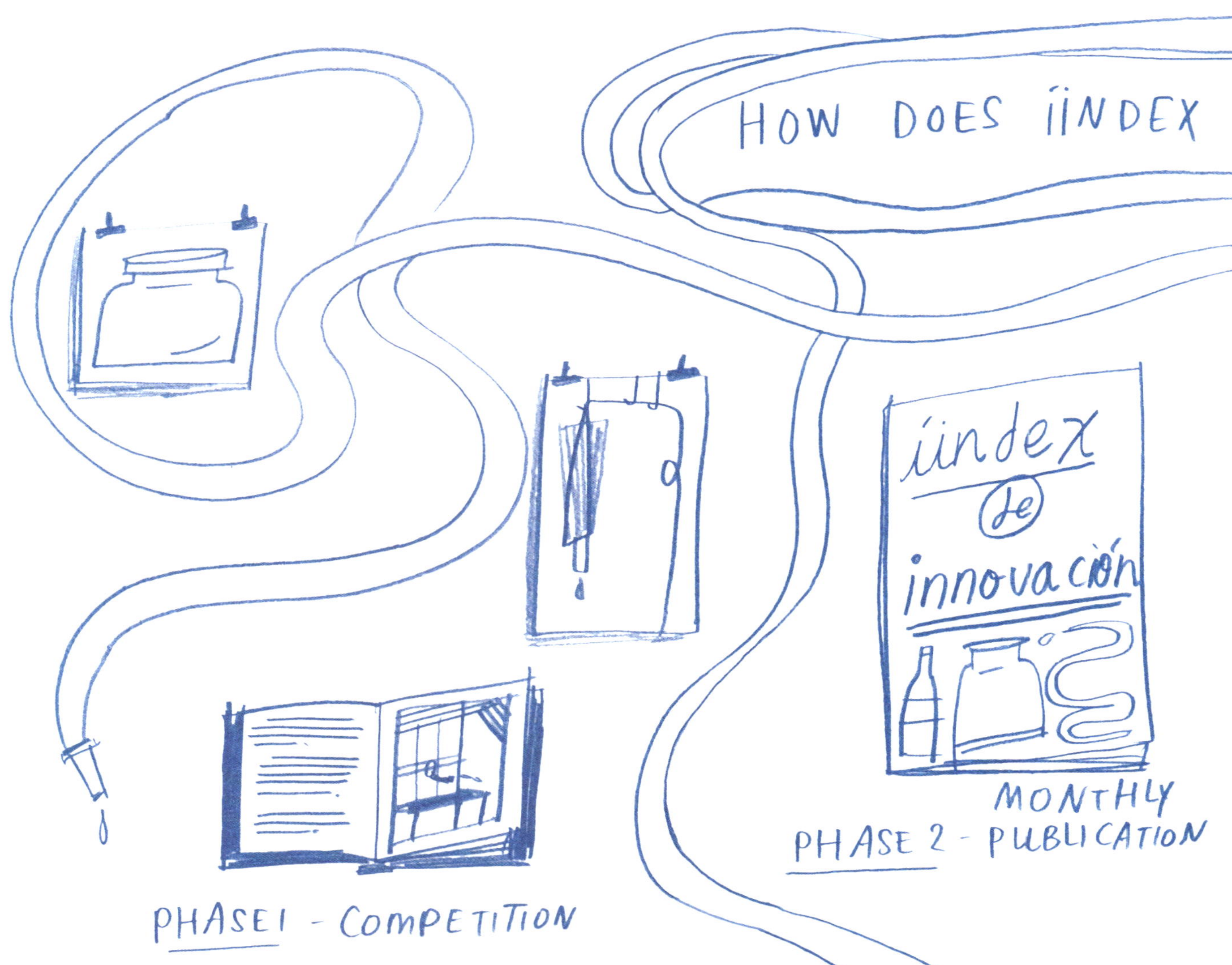

TO KICK OFF THE *ÏINDEX*, an innovation competition (similar to the X-prize) will inspire people, build community, and gather ideas from campamentos across Chile and Latin America. The winning ideas will be showcased in media such as exhibitions, publications, and posters. The competition empowers campamento families to view their everyday solutions as worthy of sharing and publishing. At the same time, the competition harnesses the attention of designers, engineers, and universities to focus on specific challenges faced by poor people around the world.

A MONTHLY PUBLICATION, circulated to campamentos across Chile, publicizes people's ideas generated by the contest, along with tips, how-to manuals, and "Innovator-of-the-month" profiles.

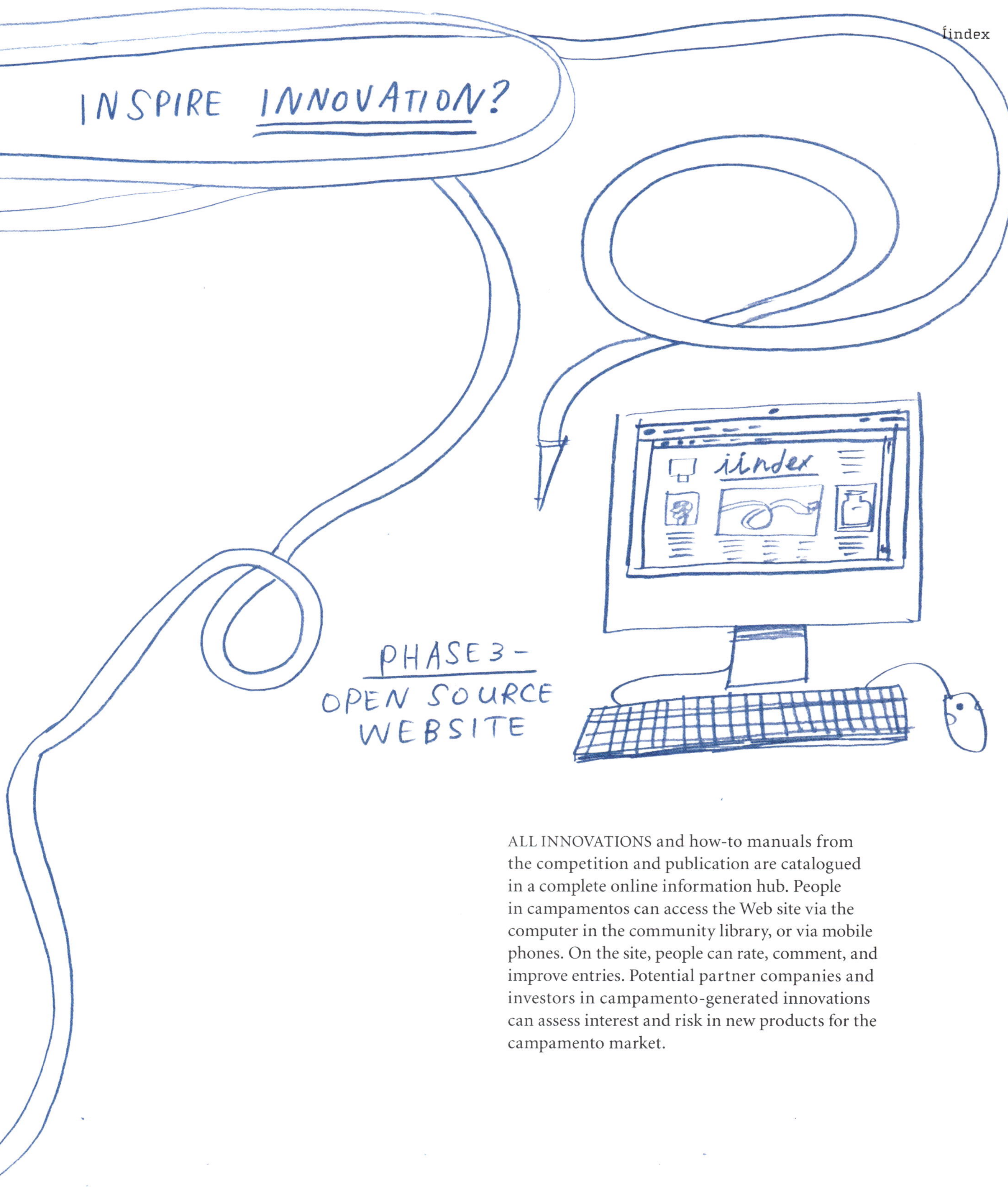

ALL INNOVATIONS and how-to manuals from the competition and publication are catalogued in a complete online information hub. People in campamentos can access the Web site via the computer in the community library, or via mobile phones. On the site, people can rate, comment, and improve entries. Potential partner companies and investors in campamento-generated innovations can assess interest and risk in new products for the campamento market.

"¡Índex de Innovación facilitates social innovation BY people in Chile's campamentos FOR people in Chile's campamentos, and beyond."

PEOPLE LIVING IN THE CAMPAMENTOS come up with innovative solutions to everyday problems, using and reusing minimal resources in ingenious ways. Ideas range from decorative table and wall coverings made from re-purposed laminated posters, to functional taps made from coke bottles, to efficient storage systems made from scrap wood. Some of these DIY solutions are shared with immediate neighbors, but most people are not aware of the useful inventions that their community or campamento dwellers in other parts of the country have created to solve daily challenges.

CLOCKWISE FROM OPPOSITE PAGE: An ingenious, whimsical tree house built by a 12-year-old boy in a less-developed campamento on the outskirts of Santiago. An invention hanging on the wall in a home in Campamento San José. A barbeque made from half a metal oil drum.

"Truly effective and sustainable social impacts must consist of solutions that will both empower the people and have market viability."

WILLIAM TANG, PRODUCT DESIGN STUDENT

THE *ÍINDEX* COMPETITION, monthly publication, and Web site advance Un Techo's *mínimo* philosophy — that people living with minimal resources can drive design innovation. C.K. Prahalad, the eminent thought leader who was influential in inspiring Un Techo's *mínimo* philosophy, cites not just the poor's participation in local problem solving, but also in worldwide economic growth: "Four billion poor can be the engine of the next round of global trade and prosperity. They can be a source of innovations… in technology, products and services, and business models," and result in "sustainable win-win scenarios where the poor are actively engaged and, at the same time, the companies providing products and services to them are profitable."

Campaign poster for the innovation competition.

Jarosita / One day, Father Felipe was sitting at his desk, telling a group of us about his interest in stars, planets, rockets, and the many scientific projects that NASA was conducting. Suddenly, he stood up from his chair and picked up a piece of rock to show us. It was about 10 inches around, and he explained that the name of this rock was "Jarosita." He said it was a special geological formation that exists only in places where there is water, and that this particular rock was recently found on Mars during one of the NASA missions to that planet. It provided evidence that there was once water on Mars — and therefore, possibly, life. / Then Father Felipe had an idea. "Why don't I give you each a piece of Jarosita," he said, "so you can take it with you and tell this story to other people?" Then he grabbed a hammer that he had hidden under his desk. We were all a bit surprised to see a Catholic Priest who keeps a hammer under his desk, as this wasn't very normal to us. But he put his rock on the floor, took his hammer and started striking the rock until he obtained three perfect pieces of Jarosita stone. / Julián and Askan, the guys from *Un Techo para Chile* who were there with us could not believe what was happening, and thought it was hilarious. "I haven't laughed this much," said one of them, "since I was in high school!" But after the successful operation, Felipe gave Penny, Dan, and me each one piece of the rock. Neither Julián nor Askan received a piece of Jarosita for themselves that day. This made us feel as though Felipe must really have liked us. After this charming meeting, we said our goodbyes and left, each carrying a piece of rock and thinking, this may be a little heavy to bring back to L.A. in our suitcase! Before departing from Chile, Dan decided to split his piece of Jarosita in three, and he gave one piece each to Julián and Askan in gratitude for introducing us to Father Felipe.

By **Felipe Berríos**, Founder of *Un Techo para Chile*,
as told to **Liliana Becera**, Faculty, Product Design,
Art Center College of Design

LOOKING INTO THE FUTURE, we realize that the new forces pushing the world forward are driven by the collaboration of entire teams of people from diverse professions and cultural and economic backgrounds who share the same passion to achieve a better world — one without poverty.

Leaving Our
Comfort Zone

By Rafael Achondo

ACCORDING TO THE WORLD BANK, there are 200 million people living in poverty in Latin America, and 80 million who survive on less than two dollars a day. Studies and statistics show a worrying lack of access to education, a shockingly high incidence of violence, and the haphazard manner in which thousands of communities struggle to survive on our continent. Taken together, these findings reveal society's inability to agree on — or disinterest in finding — concrete solutions for the inequality that we have generated.

Many theories about solving these multiple crises aim at measures and solutions that seem simple and straightforward (although they are often extremely paternalistic, and based on our own faulty perceptions rather than a true understanding of those in need). We can point the finger at government for its lack of management and deficient public policies. We can accuse the private sector for their inequitable concentration of resources. Or we can shrug our shoulders and blame society in general for not being able to create a development model that reaches everyone.

But in espousing these "simple" solutions that lie in the hands of others, we — comfortable with our own quality of life — wait for a "Super Government" to step up with radical and effective measures that create stability for families who have lived in slums and *favelas*. We wait for a nonprofit organization to connect these families to formal networks and help them organize their communities to overcome poverty. We wait, and the years go by with nothing being done.

It is not utopian to think that we as a society can achieve broad social inclusion, but to do so will require a commitment from our younger generations. Receiving a quality education is a privilege that millions of children have never even considered possible, and one that should come with

a great sense of responsibility toward those who are not so blessed. This may take many out of their comfort zones, but it is a call to action that must not go unheeded.

We cannot separate ourselves from the social necessities of our countries. The inequality and exclusion seems incomprehensible when one considers the human capacity to organize and develop solutions to scientific and technological obstacles, yet millions of children continue starving to death and entire families practically live in the trash. Brilliant minds abound to create and discover new needs, but we are seemingly incapable of ending the suffering of people whose problems are so serious that it should keep us from sleeping at night, and instead be continuously seeking solutions.

It gives me hope to think of projects such as Safe Agua, with all these trained minds working together in a grassroots effort to support and empower needy communities that have been waiting many years for a simple helping hand. Can such pioneering efforts eventually result in engineers, architects, lawyers, media professionals, designers, psychologists, technicians and professionals from different areas discussing with real families their daily obstacles, in permanent interdisciplinary workshops? Can such advances seek out leaders inside the groups, and empower them toward solutions that directly improve their quality of life? One can only hope that the answer is yes.

We need the next generations — those who will lead the planet in the years ahead — to become profoundly knowledgeable about the social needs of those who struggle to survive so that when the time comes to make decisions, the right decisions will be made. Simply put, we need to have the best human capital invested in finding the solutions that really matter. To accomplish this, it is not enough to study poverty or analyze statistics. It is necessary to get personally involved; to learn the names and see the faces of people who do not seek effortless gifts of charity, but rather require a working team that supports and open doors for them. We do not need to work on their behalf. We need to work alongside them, finding tools that empower them to become the drivers of their own development.

Safe Agua, an effort developed by *Un Techo para mi País* and Art Center College of Design, demonstrates the urgency and efficacy of involving professionals and university students in social inclusion initiatives. Yes, these students researched the 20,000 families in Chile who live in poverty, imagined what it meant to live in a space of 18 square meters, and tried to prepare themselves for an environment that they knew would be overwhelming. But when they put their feet on the ground they learned the difference between academic theory and the reality in which these people live. They put themselves in an uncomfortable position that allowed them to see a problem not through data, but through flesh and blood, and it was this revelatory experience that allowed them to conceive and develop effective solutions.

When we sought to work with Art Center, we sought to have world-class, technically and artistically proficient design teams serve those who really need design solutions. Yet the results of this project have not been so outstanding simply because of the talented people involved, but rather because of the emotional ties that were created when the students decided to put themselves in the shoes of others — experiencing their difficulties and fears — and seek solutions based on that reality.

Students, such as those who undertook the Safe Agua project, are the societal muscles that will eventually stop the world from being knocked down by inequality and exclusion. They are the reason that optimists can see the glass as half-full, and the motivation that others need to finally make a decision to take charge and heal our social wounds. The invitation to leave one's comfort zone is open for those who seek to develop in themselves and in society something that is really worthwhile; for those who care about the forgotten children and families; and for those who understand that change requires action. The future depends on us, and on the roles we are willing to assume.

Rafael Achondo is Director of Development at *Un Techo para mi País.*

Father Felipe Berríos' office, where an inflatable astronaut with Father Felipe's face on it displays his interest in science.

An Unresolved Challenge

By Father Felipe Berríos

IN THE EARLY SIXTIES, at only 43 years of age, John F. Kennedy became the president of the United States. At the beginning of his term, the courageous and charismatic leader challenged his nation and the whole world by assuring that his country would develop scientifically and technically in such a way that, before the end of the decade, no child in the world would go to bed hungry and that, through this development, man would reach the moon.

Despite the prevailing optimism of the sixties, the entire world reacted skeptically to both challenges, especially to that of reaching the moon. Thinking that man would develop scientifically and technically to such an extent as to end worldwide hunger "in our lifetime," according the words of Kennedy, was an attainable dream. There were reasons to believe him. But developing scientifically and technically in such a short period of time as to be able to "take on the goal of sending a man to the moon and getting him back to Earth safe and sound before the end of the decade," as proposed by the young president, seemed like sheer madness.

However, on July 20th, 1969, mankind witnessed the moment Neil Armstrong set foot on the moon. No human being had ever before observed the Earth from another heavenly body. In the years that followed, several costly and successful space missions took many other astronauts to the surface of the moon. They traveled across vast extensions of it in vehicles, and using various scientific devices they analyzed the natural satellite in detail and brought back to Earth many kilograms of moon rock samples.

If our science and technology was so developed, we could ask ourselves: What happened to the second part of the challenge, that of putting an end to hunger? Ending poverty and hunger is not impossible. Today more than ever before in the history of mankind man has the means to eradicate them, as pointed out by the Vatican Council II: "The human race has never enjoyed so much prosperity and richness of resources and economic power. And yet an enormous percentage of world citizens are struck by hunger and poverty…."

The problem of poverty and hunger in Chile and in the entire world is not impossible to overcome. The scientific, technical and economic conditions are there to put an end to it. But the moral consciousness and political will to do it are lacking.

Today, 40 years later, we have realized that working alone and independently does not always bring about success. The new forces pushing the world forward are based on the collaboration of entire teams made up of people from different professions and cultural and economic backgrounds who share the same passion to help eradicate poverty. In this sense, the contributions offered by the field of design are uniquely valuable and essential.

One sunny spring day, my cousins and I were playing like all the other kids in the campamento. Not many people would like to live or raise their children here, but we make the best of it. There is a big tree, the mayten, which we like for its flowers and scent. As we played beneath it, my cousin Alondra told us of her friend, Zara, a bird who lived in the tree, and who was saddened because she could not find her father. / We climbed the tree and saw a beautiful nest where there were three birds: the mom, the daughter — Zara — and the son. Zara asked us if we could help in finding her father. "A hunter called Silvestre has kidnapped him, and he lives in the city of wood." / She described this "city of wood" that lay beneath the tree, "where they say it's all wood, cardboard and nylon. At night, cold gets through, and in winter it rains inside the house, and in summer the heat and the earth are felt everywhere, and they live on some hills full of garbage." / Her words were true, although we preferred to imagine our walls in the city of wood as being surrounded by green grass with swings and flowers. / Silvestre lived in site 13, and had more than one bird in a cage. Zara described her father as a big, coffee bird with a white beak, and said that his leg was hurt when he was stolen. / "We will help you," said my cousin Cecilia. /

We went to Silvestre and asked, "How much is that big bird with a bandaged leg?" / "He is not for sale, and I don't think he would be within your reach, anyway," he answered rudely. / That was when I remembered I had seen that bird in a schoolbook that said he was protected by the fauna act. / We called the authorities, and they came right away to arrest Silvestre, and free all his birds. We left running, with Zara's father, to tell her the news. When we arrived they sang for joy. We all celebrated the reunion of the family, because family is the most important thing. / I look at my city of wood with nostalgia, but also with excitement and joy, because I make out my city of concrete in the horizon. Here, where I will not feel the cold and loneliness of my city of wood, but the warmth and love of a home, together with my family, my cousins and my favorite tree: the mayten.

Janzita Munoz is 11 years old, and lives with her family in Campamento Juntas Podemos in Lo Espejo, Santiago. Her story was selected as the winner from among 200 entries in the inaugural children's "Stories from the Campamentos" competition, organized in 2009 by Un Techo and Father Felipe Berríos.

Ramon Coronado / After working with Un Techo, I was opened to a whole new world of opportunities in design. Working with the families and seeing how much of an impact we made in those two weeks was extremely rewarding.

Elizabeth Bayne / Many assume that poverty is due to a lack of motivation or skills, but what I saw in the campamentos was an unexpected resilience and resourcefulness. Though many of the residents may not be employed, everyone worked. When formal employment was not an option due to circumstance, health, or physical conditions, entrepreneurship was key to survival. / The majority of adults in the campamentos left for work every day. If they weren't going to a formal job in town or a construction job in a new development, they were salvaging parts or precious metals from abandoned lots, doing laundry for their neighbors, selling small crafts, collecting recyclables and used clothing, or buying fruit and vegetables to resell. They created ways to support their families, took pride in their work, and were reluctant to take handouts.

Stella Hernandez / During a visit to one of the less developed areas in the outskirts of Santiago, we became intrigued by a tree house. Minutes later, we were listening to a mother of three kids speaking proudly of the artistic talent of her 12-year-old son, who had built it. / The structure and details of the construction were amazing. The builder's creativity and resourcefulness was evident in the materials he used, from the refrigerator door that made the main entrance to the windows with glow-in-the-dark stars. A little desk with toys and books, a white board to draw, and a comfortable chair combined to make a magic space above the tough reality. In a corner of the tree house I found one of the boy's most intriguing and inspiring creations: a water drinking system made with reused materials.

Stephanie Stalker / Carlos, an 11-year-old autistic child, was warm and accepting of three complete strangers who came into his home asking many personal questions in a language he couldn't understand. But the language gap was not a barrier to communication, and before long Carlos was engaged in a heated table-top World Cup soccer game with my colleague, K.C., on a field that he had drawn with colored pencils. / This interaction between two people who don't speak each other's language, come from different cultures, countries, age groups, social status, and family structures was incredibly touching.

This jarosita, a rock that is only
found in places with water, was given to
Dan Gottlieb as a gift from Father Berríos.

The flag of Chile, a gift to the
students, is "a symbol of hope,"
said Rosita Reyes. "Because this
is a community with dreams."

PLACES

CAMPAMENTO

Slum. An irregular occupation of land by eight or more families, lacking access to a basic service, like water, electricity and municipal plumbing. A densely populated, run-down, squalid part of a city, inhabited by impoverished people. 80 million people in Latin America live in campamentos, also referred to as *Villas Miseria*, *Favelas*, or *Asentamientos*.

CAMPAMENTO SAN JOSÉ

Main slum in Santiago, Chile where the field research and product testing for this project took place. Occupied by 20 families, living in *media aguas* constructed by Un Techo's volunteers.

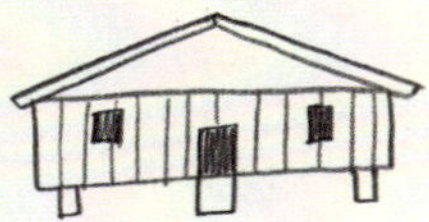

MEDIA AGUA

Literally, "shack." *media agua* is the name given in Chile to emergency homes provided by the foundation *Un Techo para Chile*. These prefabricated wooden structures provide social housing for people who are either homeless or who have lost their homes in a natural disaster. A typical *media agua* houses a family of four in 18,3 m² of interior space (6.1 m x 3 m) — smaller than a typical parking space.

CASA DEFINITIVA

Literally means, permanent housing. Defined also by the Innovation Center of UTPC as an "engine to overcome poverty"; this type of social housing provides a family with a life time living solution. *Casas definitivas* are located in networked areas of the city and aid in eliminating the current housing deficit in developing countries. Families living in the campamento save up to contribute approximately $400 to the $20,000-$30,000 government-subsidized cost of constructing the home, which they will own.

INFOCAP

Name of the University of the Worker in Chile with the goal of improving technical development skills.

CASITA

Literally means, little house — where the Art Center students stayed, on the Infocap campus.

PEOPLE AND ORGANIZATIONS

TECHO

Roof. The external upper covering of a house or other building. The word techo was adopted as the name of the organization due to its literal and metaphorical meaning connected to creating solutions to help people overcome poverty.

UN TECHO PARA CHILE

Literally means, A Roof For Chile, also known as

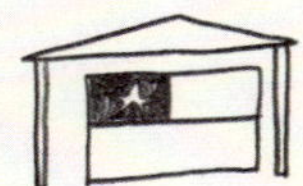

UTPC. A Chilean nongovernmental organization (NGO) whose goal is to provide home solutions for homeless people in Chile. The foundation encompasses the construction of emergency homes and social housing solutions including: micro-credits, education, libraries, and training. It was founded in 1997 by college students led by the Jesuit priest Felipe Berríos.

UN TECHO PARA MI PAIS

Literally means, A Roof For My Country, also

known as UTPMP. In 2001, *Un Techo para Chile* became *Un Techo para mi País*, which is present today in Argentina, Bolivia, Brasil, Chile, Colombia, Costa Rica, Ecuador, El Salvador, Guatemala, Haiti, Mexico, Nicaragua, Paraguay, Peru, Dominican Republic, and Uruguay.

CENTRO DE INNOVACION

Literally means, Innovation Center. A department from the foundation *Un Techo para Chile* defined by its director Julián Ugarte as the "NASA for the poor."

ROSITA

Rosa Reyes. Inspiring friend, community leader and head of Campamento San José.

THINGS AND EVENTS

OLLA COMUN

Community organization typical of campamentos involving a kitchen and the preparation of meals, where each of the residents contributes either with food, cooking expertise, or utensils.

JAROSITA

A yellowish or brownish mineral, a hydrous sulfate of potassium and iron, occurring in small crystals or large masses.

DUCHA

Shower. Also called shower bath. A bath in which water is sprayed on the body, usually from an overhead perforated nozzle (showerhead).

LLAVE

Faucet. Any device that can open and close to control the flow of liquid from a pipe.

WASHER / SPINNER

Types of machines used for washing clothing. A washer wets, soaps, and rinses the clothes. A spinner rotates to evacuate excess water out of the clothes, before they are hung to dry. Women typically wash clothes by hand, or may own one or both of these machines, which tend to break down.

A WINDOW INTO THE LIFE OF THE FAMILIES OF CAMPAMENTO SAN JOSÉ

LAURA'S FAMILY

"I'm embarrassed to have family or friends come visit because of my living situation."

FAMILY MEMBERS: Total of 8 family members, four in each *media agua*.

OCCUPATIONS: *Laura:* Student, Electricity Degree, *Paricio:* Train track construction, *Grandma:* Feria, recycles electronics, *Grandpa:* Feria, recycles electronics, *Brother:* Works for the city of Chile.

FAMILY INCOME: 440,000 Chilean pesos/month > (approx $825 per month for a family of 8).

AGE: Laura: 27 Pato: 31 Sabrina: 4

ASPIRATIONS: Laura wants to continue her education and pursue a career that deals with electricity. She has a passion for building and designing inventive things to make her life easier around the house.

PRIOR RESIDENCE: Laura lived her first 15 years with her mom. The whole family moved from the old community that was located over the mountain. The old area is bad and very dangerous.

EXISTING RESIDENCE: Campamento Fundo San José, for roughly a year.

FUTURE RESIDENCE: Definitive house. Mother does not want to move, she likes her lifestyle and her community. Laura must convince her.

CECILIA'S FAMILY

"I stopped going to school at age 10 to take care of my brother. As a child I dreamed of having a family... now I want a house and a better life for my daughter."

AGE: 45 years old

FAMILY MEMBERS: 13-year-old daughter: *Carla Andrea.* 20-year-old married son. 18-year-old son living with grandmother. Restraining order against husband with 1-day-a-week visitation, but relationship improving. A sister who lives nearby.

WORK: Collects and sells copper and other metals for money in emergencies. Has hyperglycemia, causes fluid build-up in her legs—leg pain prevents her from working outside the home.

PERSONAL HISTORY & ASPIRATIONS: Quit school at age 10 to take care of brother, because father died and mom had to work. Can't read/write, but is giving her daughter the opportunity for an education to become somebody. Enjoys nature and caring for plants.

MARIA'S FAMILY

"I saved enough money to buy this wheelchair in case one day I can no longer walk."

AGE: 58 years old

MARITAL STATUS: Married. Husband works as a security guard all day.

OCCUPATION/INCOME: Helps untangle airplane headphones for 30 pesos per untangled headphone.

MEDICAL CONDITIONS: Severe Osteoporosis. Heart Problems. High blood pressure.

FOR FUN: Watch Soap Operas

LIFE STORY: She is the mother of three. She lives next door to her middle daughter, Maria, and her grandson, Carlitos. Her older daughter (who brings her the headphone work) lives separately; her son is in jail for a few more years, for repeating a crime. She's very proud of older grandchild who is an engineer.

ASPIRATIONS: To live in her own house and for her and her family to have a decent place to live.

PAST AND FUTURE: Lived in a soccer field as watchperson until the whole family got kicked out. They had running water before. She went from living a more civilized life to life in the campamentos. She gets very sad when she thinks about her limited mobility and how soon she will have to rely on a wheelchair that she is saving for when she needs it.

FAMILY: Maria lives with her 32-year-old daughter, Maria Bernarda Penalosa. A single mother, Maria was divorced a few years ago from the father of her son. Maria used to work as a sales person at a mall and is currently unemployed. Maria has been diagnosed with depression, and suffered a miscarriage a few months ago.

Maria Penalosa aspires to live in her own house and she wishes to provide the best for her son. She wishes to learn a trade of some sort to someday open her own business.

Maria's 11-year-old son Carlitos has a lot of trouble with school due to his autism. Kids bully him and his condition is not severe enough for him to attend special classes. He has his own cell phone that his dad bought him and collects video cards. He is a soccer fanatic and can recall names and dates perfectly. He aspires to live in a house with his mother and grandmother. He still does not know what he wants to be when he grows up.

"Moving to the campamentos was very difficult and humiliating but I learned how to adjust little by little, day by day." –Maria Bernarda Penalosa, Maria's Daughter

MIREYA'S FAMILY

"Anything can be replaced...except family. [If I had to flee my home in a hurry] I would also take the bottle [that my niece made]. But if I did have time, I would take things I use more often, like my washer and spinner."

AGES / OCCUPATIONS: *Mireya*, 54 *Christopher*, son: 14 – Student, watches and washes cars at the feria. *Jonathan*, son, 17 – Student, watches and washes cars at the feria. *Ermernio*, husband, 58 – Unemployed, sells vegetables door to door. *Fernanda*, daughter.

INCOME: Disability (for asthma): 75,000 / mo. Disability (for Christopher): 52,000 / mo. Vegetable sales: 48,000 / mo.

PERSONAL HISTORY & ASPIRATIONS: As a child, Mireya lived in the country, in the south of Chile. She used to live in the "old campamentos." She wants permanent housing for warmth, and wants a real bathroom that doesn't smell.

LUIS

"I leave my door open because I like to see people passing by and have the sunlight come in."

MARITAL STATUS: Divorced

OCCUPATION: Unemployed due to health reasons (formerly a farmer). Receives a pension.

PLANS & ASPIRATIONS: Wants to paint his house yellow. Wants to go back to work. A second *media agua* in his back yard. Wants to visit his cousin in Washington. He plans to move to the *casa definitiva* with his sister and her family; as a single person does not qualify for the definitive housing.

WHAT HE VALUES: Visits from friends and family. His new oven (good brand that he can trust).

ACKNOWLEDGEMENTS

This publication is a collaborative effort, and would not have been possible without the talent, shared experiences, and dedication of the young designers and committed professionals from Art Center College of Design and *Un Techo para mi País*, who came together to tell the story of the Safe Agua project. We are grateful to all for their participation, and remain inspired by the opportunity to have partnered with the families of Campamento San José to design a better tomorrow.

UN TECHO PARA MI PAÍS / UN TECHO PARA CHILE

Rafael Achondo
Director of Development,
Un Techo para mi País

Felipe Berríos
Founder,
Un Techo para mi País

Juan Pedro Pinochet
Executive Director,
Un Techo para mi País

Andrés Iriondo
Centro de Innovación,
Un Techo para Chile

Askan Straume
Centro de Innovación,
Un Techo para Chile

Julián Ugarte
Director,
Centro de Innovación,
Un Techo para Chile

CAMPAMENTO SAN JOSÉ RESIDENTS

Rosita Reyes
Campamento Head and Coordinator,
También Somos Chilenos Corporation

Aguilera Orellana Family
Clavijo Escobar Family
Escobar Clavijo Family
Mesa Opazo Family
Miranda Quiróz Family
Miranda Sepulveda Family
Muñoz Godoy Family
Nieves Cuevas Family
Ojeda Santibañez Family
Peñaloza Abarca Family
Peñaloza Meza Family
Poblete Garcés Family
Rivas Elocudo Family
Salinas Reyes Family
Salinas Ulloa Family
Santibañes Millar Family
Luis González
Julio Moreira
Luis Reyes and Rodolfo Reyes
Mario Soto

SPECIAL THANK YOU

Carlos Hinrichsen

Former ICSID President and Director, School of Design, Instituto Profesional DuocUC of the Pontificia Universidad Católica de Chile, who introduced *Un Techo para Chile* to Designmatters at Art Center.

ART CENTER COLLEGE OF DESIGN

Dr. Lorne M. Buchman
President

DESIGNMATTERS DEPARTMENT

Mariana Amatullo
Vice President and Director

Elisa Ruffino
Producer and Senior Associate Director

SAFE AGUA STUDIO

Department Chairs

David Mocarski
Environmental Design

Karen Hofmann
Product Design

Faculty

Dan Gottlieb
Environmental Design

Penny Herscovitch
Environmental Design

Liliana Becerra
Product Design

Students

Elizabeth Bayne
Graduate Broadcast Cinema

Jacqueline Black
Product Design

K.C. Cho
Product Design

Ramon Coronado
Graphic Design

Narbeh Dereghishian
Product Design

Stella Hernandez
Environmental Design

Erica Li
Environmental Design

Nubia Mercado
Transportation Design

Stephanie Stalker
Environmental Design

William Tang
Product Design

Diane Jie Wei
Product Design

Jessica Yeh
Environmental Design

SAFE AGUA MOTION DESIGN

Department Chair

Nikolaus Hafermaas
Graphic Design, and
Acting Chief Academic Officer

Faculty

Ming Tai
Graphic Design

Students

Ian Abando
Illustration

Gürkan Erdenli
Graphic Design

James Kim
Graphic Design

Micael Mazza
Graphic Design

ChiaChi "Nadia" Tzuo
Graphic Design

Jason Yey
Graphic Design

GUEST PARTICIPANTS

Adlai Wertman
Founding Director,
Society and Business Lab,
Marshall School of Business,
University of Southern California

Abby Fifer Mandell
Director of Education,
Society and Business Lab,
Marshall School of Business,
University of Southern California

Patrice Martin
Social Innovation Lead
and Systems Designer,
IDEO

Dirk-Mario Boltz
Marketing Professor,
HWR Berlin School of Economics and Law

Viviana Araneda
Trade Commissioner,
ProChile Chilean Trade Commission,
Chilean Ministry of Foreign Affairs

Nicolo Gligo
Executive Director,
Chilean Economic Development Agency

Cristóbal Barros
Technical Secretary,
Directorate of Energy, Science,
Technology and Innovation,
Chilean Ministry of Foreign Affairs

Karen Molina
Coordinator of Asia Pacific and Human
Capital Development, Directorate of Energy,
Science Technology and Innovation,
Chilean Ministry of Foreign Affairs

Lilian Rodriguez Walker
Assistant Trade Commissioner
ProChile Chilean Trade Commission,
Ministry of Foreign Affairs

Lilina Mejias
Assistant Trade Commissioner,
ProChile Chilean Trade Commission,
Ministry of Foreign Affairs

ADDITIONAL IMAGE CREDITS

Ramon Coronado: Cover typography (digitally illustrated by Eric Mathias) / Pei-Jeane Chen: Table of Contents collage, page 5
Evangeline Joo: pp. 1, 13, 32, 33, 34, 69, 137, 141, 147 / Giancarlo Llacar: pp. 3, 9, 25, 32, 33, 34, 129, 151 / Lisa Wagner: pp. 3, 9, 141
Ping Zhu: End pages, pp. 21, 41, 47, 48, 73, 74, 78, 83, 84, 85, 90, 93, 94, 95, 100, 103, 104, 105, 110, 113, 114, 115, 120, 122, 123, 124, 125, 130, 131, 132, 135, 152.